Small Electric Motors

Use, Selection, Operation, Repair, and Maintenance

by Rex Miller and Mark Richard Miller

An Audel® Book

Macmillan Publishing Company
New York

Maxwell Macmillan Canada
Toronto

Maxwell Macmillan International
New York   Oxford   Singapore   Sydney

SECOND EDITION

Copyright © 1992 by Macmillan Publishing Company, a division of Macmillan, Inc.
Copyright © 1984 by The Bobbs-Merrill Co., Inc.

While every precaution has been taken in the preparation of this book, the Publisher
assumes no responsibility for errors or omissions. Neither is any liability assumed for
damages resulting from the use of the information contained herein.

Macmillan Publishing Company   Maxwell Macmillan Canada, Inc.
866 Third Avenue   1200 Eglinton Avenue East, Suite 200
New York, NY 10022   Don Mills, Ontario M3C 3N1

Macmillan Publishing Company is part of the Maxwell Communication
Group of Companies.

Production services by the Walsh Group, Yarmouth, ME.

Library of Congress Cataloging-in-Publication Data
Miller, Rex.
    Small electric motors : use, selection, operation, repair, and
  maintenance / by Rex Miller and Mark Richard Miller.
        p.   cm.
    "An Audel book."
    Includes index.
    ISBN 0-02-584975-1
    1. Electric motors, Fractional horsepower.   I. Miller, Mark
  Richard.   II. Title.
  TK2537.M533   1992
  621.46—dc20   92–8902
                                                            CIP

Macmillan books are available at special discounts for bulk purchases for sales
promotions, premiums, fund-raising, or educational use. For details, contact:

                        Special Sales Director
                        Macmillan Publishing Company
                        866 Third Avenue
                        New York, NY 10022

10 9 8 7 6 5 4 3 2 1

Printed in the United States of America

Edison called Sprague's motor—about 1885—the first practical machine to make electricity do industry's work.

Contents

Preface

Electric motors, especially those of fractional-horsepower size, play a very important role in furnishing power to the industrialized world. Robots make great use of small motors, as do automobiles, home appliances, and aircraft. Their versatility, dependability, and economy of operation cannot be equaled by any other form of motive power. It is estimated that electric motors are utilized in over 90 percent of industrial applications (and this figure would be higher except for the absence of power lines in some remote areas). The application of electric motors in the average home also has reached a high degree of utilization, ranging from the smallest units, found in electric clocks, to the larger units in air conditioners, heating plants, etc. It is a rare individual, indeed, whose daily life is not affected in some way by electric motors.

This book has been prepared to give practical guidance in the selection, maintenance, installation, operation, and repair of electric motors. Electricians, industrial maintenance personnel, and installers should find the clear descriptions and illustrations, along with the simplified explanations, a ready source of information for the many problems they encounter. Both technical and nontechnical persons who desire to gain knowledge of electric motors will benefit from the theoretical and practical coverage this book offers.

Acknowledgments

No book can be written without the aid of others. It takes a great number of persons to put together between two covers the information available about any particular technical field. The field of electric motors is no exception. Many firms have contributed to the information, illustrations, and analysis of the book.

The author would like to thank everyone involved for his or her contributions. Some of the firms that supplied technical information and illustrations are listed below:

Allis-Chalmers, Inc.
Amprobe, Inc.
Brown & Sharpe Co.
Fasco Industries, Inc.
Hayden Switch and
    Instrument, Inc.
Howard Industries, Inc.
Lennox Industries, Inc.
Northland Div. of Scott and
    Fetzer Company
Robbins & Meyers, Inc.
Square D Company

Voorlas Manufacturing Company
Weston Electric Company
Bodine Electric Co.
Doerr Electric Corporation
General Electric Company
Leeson Electric Corporation
Mechaneer, Inc.
Sears, Roebuck & Company
Stanley Tool Company
Stock Drive Products
Westinghouse Electric Corporation

PART ONE

Introduction

CHAPTER I

Motor Repair Tools

Whenever you need to repair an electric motor, you will need standard tools as well as some tools which ordinarily are not found in the home workshop. In some instances you may need to locate or devise your own holders, pullers, or winders. This chapter will deal with some of the special tools needed, as well as some of the ordinary tools found in every workshop.

Screwdrivers

Most people have screwdrivers around the house or the shop. There are two types of screwdriver blades, the standard slot type and the Phillips-head type, which has a crossed end. Screwdrivers come in thousands of variations. They may have wooden handles or they may have plastic handles; today most screwdrivers have plastic handles.

Fig. 1–1 shows a flat-blade screwdriver with a plastic handle. The plastic handle is very helpful when working around electricity. Plastic is supposed to be shockproof if the handle is kept clean. The blade tip may vary in size from ⅛ inch to ¼ inch. The shaft is from 4 to 8 inches long and is usually of nickel-plated chrome-vanadium steel. The tips or points must withstand the force applied when a screw sticks or is hard

Fig. 1–1. Flat-blade screw-driver. *(Courtesy Stanley)*

to remove. The main thing to remember when using a screwdriver is to get a good fit between the tip of the screwdriver and the slot in the screw. This will prevent damage to the screw head and the screw-driver.

The Phillips screwdrivers (Fig. 1–2) have point sizes. The No. 0 point is ⅛ inch, the No. 1 is 3⁄16 inch, No. 2 is ¼ inch, No. 3 is 5⁄16 inch, and No. 4 is ⅜ inch. In this case it is very important to make sure the right size point is used for the screw head. That is why it is best to obtain a complete set, from No. 0 to No. 4, to fit all screw heads.

Pliers

There are a number of pliers available for special jobs. The pliers in Fig. 1–3 are indicative of the variety available for work in the electrical or electronics field. Each is designed for a particular job.

1. A 4-inch midget for close work.
2. 4-inch pliers for fast, clean tip cutting. It has a tapered nose and

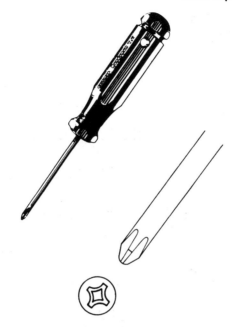

Fig. 1–2. Phillips-head screwdriver. *(Courtesy Stanley)*

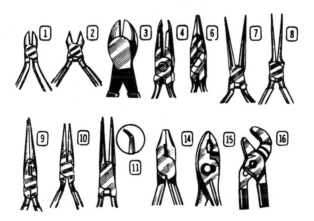

Fig. 1–3. Various types of pliers.

nearly flush cutting edges and it will cut to the tip. It produces burr-free cuts.

3. 7-inch diagonal pliers for heavy-duty cutting.

4. 4½-inch thin needle-nose pliers with cutter at the tip.

6. 5-inch thin chain-nose pliers whose smooth jaws are slightly beveled on inside edges.

7. 5½-inch pliers with the fine serrated jaws for firm gripping or looping wire.

8. Slim serrated jaws (6 inches long) permit entry in areas inaccessible for regular long-nose pliers.

9. Long-nose pliers (6½ inches) with side cutter.

10. Long-nose pliers (6½ inches) without side cutter.

11. Thin bent-nose pliers (5 inches) with fine serrated jaws; 60° angle thin bent-nose for thin wire.

14. 8-inch serrated upper and lower jaws with side cutter.

15. 8-inch chrome-plated combination pliers for general use.

16. Four-position 10-inch utility pliers with forged rib and lock design with serrated jaws.

Hammers

There are a number of types of hammers. The three types mentioned here are best suited for motor repair work. Most people who have a basement or garage workshop undoubtedly already have them.

The claw hammer is used to drive nails and to work mainly with wood; it has claws with which to pull out nails. It is the most common type of hammer, but it is used only occasionally in motor repair work, whenever the ball-peen hammer is not available. Fig. 1–4 shows the claw hammer most often found around the home.

The ball-peen hammer (Fig. 1–5) has a rounded top and a flat larger-diameter bottom surface.

In some cases a mallet is needed to force a connection between the case and the stator or to move the armature gently into place. In most instances you will use a rubber mallet, but in more demanding situations you may need a plastic-tipped mallet. These will do the job without marring the surface of the metal. You are advised to be careful with your hammer or mallet blows. Fig. 1–6 shows the plastic-tipped mallet.

Fig. 1–4. Claw hammer. *(Courtesy Stanley)*

Hacksaws

Hacksaws are very useful for cutting metal. Fig. 1–7 shows the hacksaw blade being installed and as it is used to cut metal. A blade will break if it is not properly inserted and tightened; so once the blade is pointing in the right direction—away from the handle—tighten it. Note that the cutting takes place when the hacksaw is pushed away from the operator. Lift the saw when bringing it back to starting position. Riding the blade as it is drawn back through the metal ruins it.

Blades come in 8-inch, 10-inch, and 12-inch lengths. The hacksaw is usually adjustable to fit any of the three lengths of blades. Hacksaw blades also come in a number of tooth sizes. Use 14-teeth-per-inch blades for cutting 1-inch or thicker sections of cast iron, machine steel, brass, copper, aluminum, bronze, or slate. Use the 18-teeth-per-inch blades for cutting materials $\frac{1}{4}$- to 1-inch thick in sections of annealed tool steel, high-speed steel, rail, bronze, aluminum, light structural shapes, and copper. The 24-teeth-per-inch blade is used for cutting material that is $\frac{1}{8}$- to $\frac{1}{4}$-inch thick in sections; it is usually best for iron, steel, brass, and copper tubing, wrought-iron pipe, drill rod, conduit, light structural shapes, and metal trim. The 32-teeth-per-inch blade is

Fig. 1–5. Ball-peen hammer.

Fig. 1–6. Plastic-tipped mallet. *(Courtesy Stanley)*

used for cutting material similar to that recommended for 24-tooth blades.

A hacksaw blade is handy when you want to check for a short in an armature or remove a wedge, as shown in Fig. 1–8. A power hacksaw blade is usually handy to have for removing wedges.

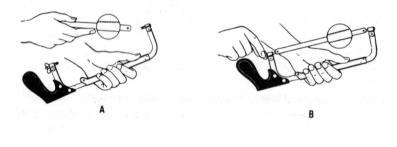

A   B

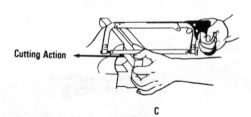

Cutting Action ⟵

C

Fig. 1–7. Inserting a hacksaw blade, and using the hacksaw.

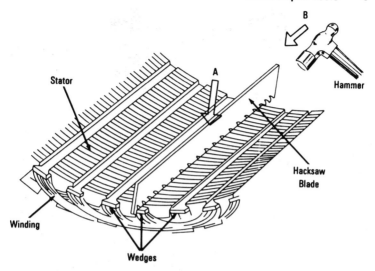

Fig. 1–8. Using a hacksaw blade in motor repair work.

Wrenches

Wrenches are used to tighten and loosen nuts and bolts. There are two general types of wrenches—adjustable and nonadjustable. Adjustable wrenches have one jaw that can be adjusted to accommodate different size nuts and bolts and may range from 4 to 18 inches in length for different types of work (Fig. 1–9). When adjustable wrenches are

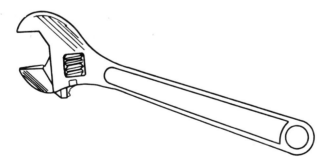

Fig. 1–9. Adjustable wrench.

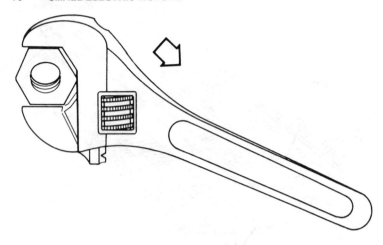

Fig. 1–10. Note the direction of force with this type of wrench.

used, there are two rules to remember: (1) Place the wrench on the nut or bolt so that the force will be placed on the fixed jaw, and (2) then tighten the adjustable jaw so that the wrench fits the nut or bolt snugly (Fig. 1–10).

Nonadjustable wrenches have fixed openings to fit nuts and bolt heads. Fig. 1–11 shows a nonadjustable open-end wrench, while Fig. 1–12 shows a nonadjustable box-end wrench. These wrenches are available in sets. They are also available in metric sizes. Openings are

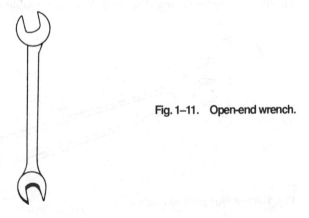

Fig. 1–11. Open-end wrench.

Fig. 1–12.   Box-end wrench.

actually 0.005 inch to 0.015 inch larger than the size marked on the wrench. This is to allow the wrench to be slipped over the nut or bolt head easily. Make sure, however, that the wrench fits the nut or bolt head properly (Fig. 1–13). If not, it is possible to damage the nut or the bolt head. It is safer to pull than to push a wrench. If you exert pressure on a wrench and the nut or bolt head suddenly breaks loose, there is a chance that you will suffer an injury to your hand. If the wrench must be pushed rather than pulled, use the palm of the hand so that the knuckles will not be injured if there is a slip.

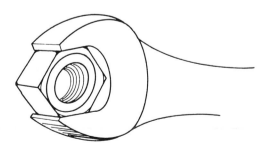

Fig. 1–13.   Proper fit of nut and wrench.

Allen Wrenches

Allen wrenches are designed to be used with headless screws which are used in many devices as setscrews (Fig. 1–14). Allen wrenches come in various sizes to fit any number of setscrews (Fig. 1–15). A complete set is especially helpful to the motor repairperson.

A newer type of recessed setscrew head has an Allen-type hole, but ridges in each flat side make it difficult for an Allen wrench to fit. Newer hex-socket key sets are made in $3/32$- to $1/4$-inch sizes, with eight blades in a set (Fig. 1–16).

Socket Wrenches

Socket wrenches may be used in some locations that are not easily accessible to the box-end wrench or the open-end wrench. Sockets are easily taken off the ratchet and replaced with another size. Sockets come in a 12-point or 6-point arrangement. Make sure you use the proper size for each nut or bolt head. This type of wrench is also available in metric sizes. Fig. 1–17 shows some of the sockets, extensions, flexible handles, and universal joints that make sockets effective in almost any location.

Torque Wrenches

Torque wrenches are made so that you can apply the proper torque to various bolts and nuts, and are made to fit various socket

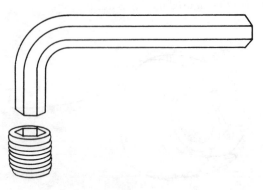

Fig. 1–14. **Allen wrench and setscrew.**

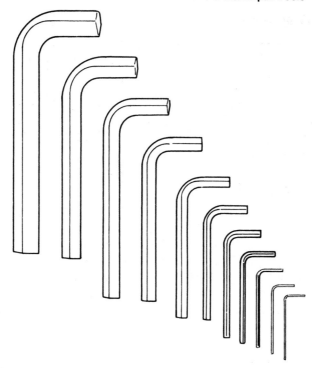

Fig. 1–15. Allen wrench set.

drives. Two popular sizes of torque wrenches are the ⅜-inch drive and the ½-inch drive. Torque wrenches are made to measure in inch-pounds and also in foot-pounds. Use the proper wrench according to the torque that has to be applied. The wrenches come in various handle lengths. Normally, the longer the wrench, the greater the torque

Fig. 1–16. Socket key set.

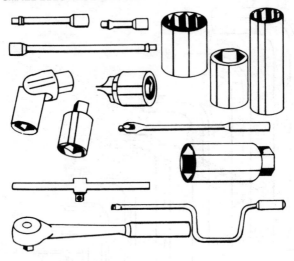

Fig. 1–17. Socket wrench set.

the wrench will measure. A typical torque wrench is shown in Fig.
1–18.

Nut Drivers

The nut driver is nothing more than a socket attached to a screwdriver handle. It is an excellent tool for most motor repair applications.
Nut drivers come in a variety of sizes and usually have the size
stamped on the plastic handle. Sometimes they are color-coded according to size. Fig. 1–19 shows a set of nut drivers.

Fig. 1–18. Torque wrench.

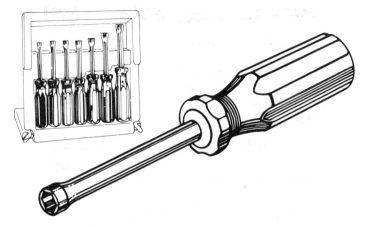

Fig. 1–19. Nut-driver set. *(Courtesy Stanley)*

Other Tools

In motor repair it is necessary, in some cases, to remove bearings from the end bell of the motor. A bearing tool (Fig. 1–20) makes the task somewhat easier. The set illustrated has nine adapters that easily and quickly remove or insert any sleeve motor bearing or bushing with a ½- to 1-inch inside diameter. It eliminates the chance of broken bearings or end bells.

Some bearing removals need a different approach. The pulley or gear puller can, in some instances, be used to remove a bearing that has stuck to the armature. These pullers come in a variety of sizes and styles (Fig. 1–21). They fit almost any application that the motor repairperson may have.

Fig. 1–20. Bearing tool with adapters.

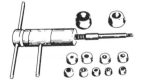

Fig. 1–21. Gear pullers.

Bushing Tools

Bushing tools have been designed for removing or inserting bushings in motors. They are handy time savers. A complete set usually consists of 20 pieces—the box, 3 drivers, and 16 adapters that cover a $\frac{3}{8}$- to $1\frac{1}{4}$-inch range (Fig. 1–22).

The solderless connector crimper (Fig. 1–23) is very useful in motor repair work. It is used to make tight connections which do not require soldering, and a good connection is necessary to withstand the vibration of a motor. A number of connectors have been designed for electrical work. The tool and a kit of connectors and lugs of various sizes are available at most electrical supply houses.

There is another type of solderless connector that does not require a special tool to apply. This is the wire nut, which makes a connection by virtue of a piece of copper coiled inside an insulated cover. By twisting the wire nut after the wires are inserted it is possible to make a good electrical connection (Fig. 1–24).

Soldering Iron

The soldering iron (Fig. 1–25) comes in handy for making a solder connection that will take the vibration and withstand corrosion. Sol-

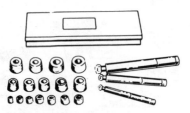

Fig. 1–22. Bushing tool set.

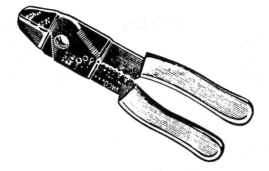

Fig. 1–23. Solderless connector crimper.

dering irons are available from about 15 watts to over 600 watts. The best all-purpose size for use in the shop is about 100 watts. This will do the job in almost all cases where larger wires are concerned. The small 15-watt irons are very useful for electronics work.

The soldering gun (Fig. 1–26) is very handy for making quick disconnects of soldered joints. It can also be used for many coldsolder joints when the person using it heats up the tip, places solder on the tip, and lets it cool on the joint. The wires being soldered or the metal surface and the wire being connected to it must be of sufficient temperature to melt the solder. This means that the gun must be left in one spot long enough to cause the joint to be heated to the temperature needed to melt the solder. The secret is to heat the material and not the solder.

In some isolated cases it may be necessary to melt silver solder or to braze a joint. A miniature torch (Fig. 1–27) welds, brazes, solders, and cuts metal with pinpoint accuracy. It uses butane gas and oxygen to produce a very hot flame in a very restricted area.

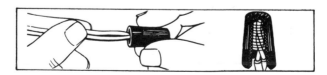

Fig. 1–24. Wire nut solderless connector.

Fig. 1–25. Soldering iron.

Wire Gages

Wire gages are needed to measure the wire used in the repair business. There are numbers on the gage (Fig. 1–28) that tell you the size of the wire. Keep in mind, however, that the insulation on the wire—in the case of FORMVAR insulation—will read one size larger on the gage than the wire. Keep in mind also that the wire is moved through the slot in the gage. The hole is there to pass the wire through. The slot does the measuring. Pull the wire free of the slot and through the hole. Decimal equivalents are usually stamped on the metal disc on the opposite side of the gage numbers. No repair shop should be without a wire gage.

Stator-Holding Stand

The stator-holding stand is a necessary item if you are going to rewind motors. Fig. 1–29 is one example of such a stand. It is self-explanatory. In most cases the stand is bolted to a table or mounted in a vise, to secure it while the work is being done.

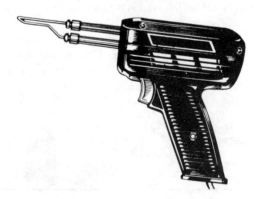

Fig. 1–26. Soldering gun.

Fig. 1–27. Gas welding
torch.

Armature-Holding Stand

An armature-holding stand is necessary to work freely on an armature. Fig. 1–30 shows how the stands are used in the process of removing the wooden or plastic wedges of the armature.

Reel Rack

The reel rack (Fig. 1–31) is necessary to hold your reel of wire so that it can be unreeled easily without getting tangles or kinked. The term used here is *de-reeled.* The tension device keeps the wire from falling on the floor or becoming tangled. A number of different types are available. This one (Fig. 1–31) illustrates only one type.

Fig. 1–28. Wire gage. *(Courtesy L. S. Starrett)*

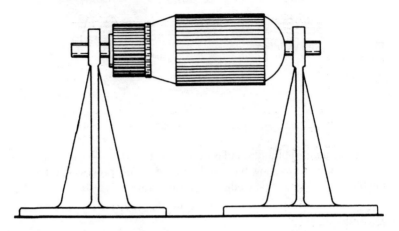

Fig. 1–29. Stator-holding stand.

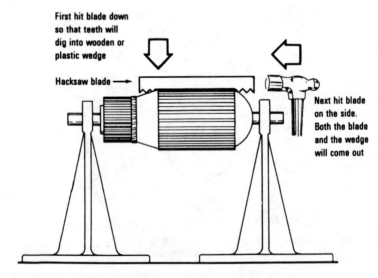

First hit blade down so that teeth will dig into wooden or plastic wedge

Hacksaw blade ➔

Next hit blade on the side. Both the blade and the wedge will come out

Fig. 1–30. Using a hacksaw to remove wedges on older motors.

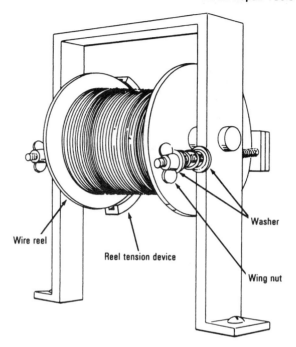

Washer

Wire reel

Reel tension device

Wing nut

Fig. 1–31. Bench setup of a reel rack.

Lathe

A lathe such as the one shown in Fig. 1–32 comes in handy for rewinding or turning down commutators. It can be used for any number of machining operations that may be needed to repair some heavily damaged motors. This lathe can be used for turning, facing, grinding, buffing, sawing, sanding, threading, milling, drilling, and coil winding. The size of the lathe will depend on your budget and applications for which it is needed.

Vise

No shop would be complete without a vise. It can be used to hold almost everything during the rewiring operation. It can be used to hold armatures, if you line the jaws of the vise with copper or some soft

Fig. 1–32. Lathe.

metal so the clamped piece is not damaged when the jaws are made tight. Vises come in almost any size you can imagine. The size you will need is determined by the size of the motors you handle most often.

Storage Cabinet

Since there are so many parts and pieces involved when working on motors, it is best to have some method of storing small parts. The drawers shown in Fig. 1–33 make an ideal storage arrangement for nuts, bolts, screws, and other small parts. Each drawer should be la-

Fig. 1–33. Small-item storage cabinet.

beled and loose parts catalogued by placing them into the proper drawer. Then, when you need a part, you will not spend time looking through everything else to find it. An organized shop can make the difference between a profitable business and a losing proposition.

Meters and Test Devices

A number of meters are available for use in testing and working with electric motors. The VOM, or volt-ohm-milliammeter (Fig. 1–34), is very useful in checking for shorts and opens.

The best way to check a meter is by its ohms-per-volt (ohms/volt) rating. The higher the ohms/volt rating, the better the meter. It should have at least 20,000 ohms/volt on DC and at least 5000 ohms/volt on AC. This should be sufficient for checking most motor troubles. A shunt ohmmeter that will measure from 0 to 200 ohms is also useful, since most motor windings and armatures are less than 300 ohms. In some cases it is possible to obtain a VOM with both a series ohmmeter for measuring high resistances and a shunt ohmmeter for measuring low resistances.

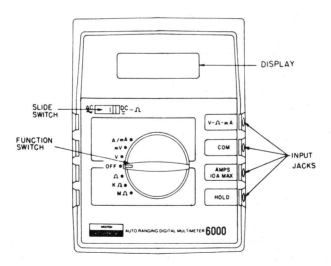

Fig. 1–34. Digital volt-ohm-ammeter. *(Courtesy Weston)*

The clamp-on type of ammeter (Fig. 1–35) can be used to measure current. It will clamp around *one* wire of an AC line to a motor and read the current being drawn by the motor. The nameplate usually tells how much current the motor will draw during normal operation.

Insulation Tester

One of the ways to measure insulation resistance is to use a megger or insulation tester (Fig. 1–36). A megger quickly locates intermittent shorts, bad electrical connections, insulation breakdowns, and conductor failures due to wear, moisture, and corrosion. Insulation resistance is calibrated directly in megohms (one million ohms). The voltage output of the megger is from 500 volts DC to 2000 volts DC, and can produce a painful shock if not handled properly. However, it is well worth its price since it can locate problems that would otherwise be most difficult or impossible to find.

Tachometer

The tachometer is a very useful device for measuring the speed of a motor. It can help locate possible troubles and can indicate if the motor is operating as it should after it has been repaired.

A hand-held tachometer is shown in Fig. 1–37. Its shaft speed will measure from 50 to 4000 rpm. It can be placed on the open end of a motor shaft, or it can be used on motors, saws, compressors, fans, pumps, grinders, and other electric motor-powered tools or equipment. A cone-shaped tip is used for shafts with center holes; a cup-

Fig. 1–35. Clamp-on type ammeter. *(Courtesy Amprobe)*

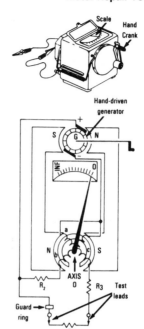

Fig. 1–36. Megger.

shaped tip is used for flat-end shafts. Other types of tachometers are available which use a strobe light to detect the number of revolutions per minute; however, these are somewhat expensive.

Micrometer

One other device should be considered as necessary for any well-equipped motor repair facility—the micrometer (Fig. 1–38). The mi-

Fig. 1–37. Tachom-eter. *(Courtesy Stew-art-Warner)*

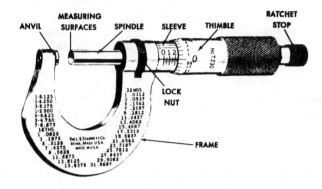

Fig. 1–38. Micrometer.

crometer is a very accurate instrument used to check the thickness of insulation, the diameter of wire, and other such applications. Instructions for using a micrometer are furnished with the instrument or can be found in almost any textbook on metal working.

This chapter has included some of the tools used in the motor repair business. You may want to buy them all at one time or acquire them slowly as they are needed. In some cases you may want to develop your own designs to fit your own habits. In any case, this is not an exhaustive listing but an attempt to show some of the tools needed to do a good job in motor repair.

CHAPTER 2

Motor Repair Supplies

A good repair shop has enough supplies on hand to avoid repeated trips to the supplier. It takes time for a new repair service to build up a complete supply of parts. However, there are commonly used parts that should always be in stock. This chapter deals with the motor repair supplies most often asked for.

Setscrews

Setscrews are used to fasten the pulley to the motor shaft in electric motors. An Allen wrench will remove or loosen the setscrew sufficiently to remove the pulley. Care should always be exercised in loosening a setscrew because if loosened too much, it may drop out and be lost as the pulley is removed from the shaft of the motor.

Setscrews should be a standard item in a shop's supplies. They can be purchased in 80-piece sets. The sets come in a compartmentalized box in various lengths of $\frac{1}{4}$-20, $\frac{5}{16}$-18, and $\frac{3}{8}$-16 thread. Each size of screw should be kept in its proper section of the storage box. Several forms of setscrews are shown in Fig. 2–1.

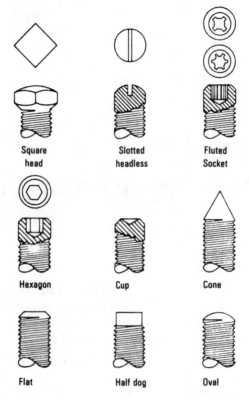

Fig. 2–1. Types of setscrews.

Carbon Brushes

Carbon brushes connect the revolving armature to the external circuit of a motor. The brushes can be square, rectangular, or round (Fig. 2–2). They are usually sold as a part of a kit.

Electric motors that use portable tool-type carbon brushes include drills, buffers, saws, sanders, and grinders. Differently shaped brushes are needed for other kinds of electric appliance motors. These include motors for fans, mixers, and AC/DC motors. Along with a plentiful supply of brush kits, a large number of springs should also be available.

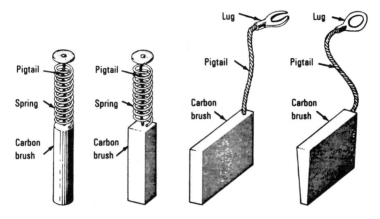

Fig. 2–2.   Carbon brushes for electric motors.

There are always many uses for a variety of different size springs (Fig. 2–3).

Bushings and Bearings

Bronze bushings make up bearings for some motors. They are easily damaged by over-tightening a belt. This puts undue stress on one side of a bearing surface. The bushing or bearing can wear and cause the armature to strike the field. This, in turn, will damage the armature and the field pole.

Bronze bushings, which are used as bearings, can be purchased in

Fig. 2–3.   Assortment of various-size springs. *(Courtesy General Electric)*

56-piece kits. The kits are made of styrene and have sections that keep the various sizes separate. There are nine different sizes of bushings in a kit. They range from ⅜-inch ID × ½-inch OD to ¾-inch ID × 1-inch OD.

Lack of lubrication is a major cause of wear in sleeve bearings. Lubricants used with sleeve bearings serve a different purpose from those used for ball bearings. The lubricant must actually provide an oil film that completely separates the bearing surface from the rotating shaft member. Thus metal-to-metal contact is eliminated. The oil film is automatically formed when the shaft begins to turn, and it is sustained by the motion of the shaft. The rotational force of motion sets up pressure in the oil film wedge that supports the load. The wedge-shaped film of oil is absolutely essential for effective sleeve-bearing lubrication as it prevents the destruction of the bearing. If the highest-quality oil is used, there will be longer wear and less trouble from the motor. On the other hand, over-lubrication should be avoided since the oil may enter the centrifugal switch and cause shorting. Also, excess oil always attracts dust and dirt.

It should be noted that the bearing also serves as a guide for shaft alignment (Fig. 2–4). Consequently, a replacement bearing must be an exact replacement, not just an approximate size.

Sleeve bearings are less sensitive to a limited amount of abrasive or foreign material than are ball bearings. They have the ability to absorb small, hard particles into the soft undersurface of the bearing. However, good maintenance practice demands that the bearings be kept clean. In small motors, dirty oil or insufficient lubrication can add enough friction to cause the bearings to seize or freeze in place.

A good practice is to replace the oil every six months. If the motor is operated under very dirty conditions, it should be cleaned more often. Keep in mind that sleeve-bearing motors tend to lose the oil film when stored for one year or more.

Shaft Adapters

Shaft adapters are used to convert the shaft of a motor to a more usable size. The shaft adapter shown in Fig. 2–5 has an Allen-head setscrew to secure the adapter to a threaded motor shaft. The illustrated adapter has a ¼-inch diameter and is 1 inch long.

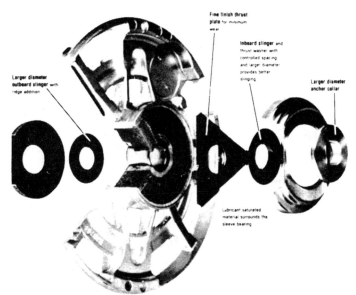

Fig. 2–4. Method of lubricating a bronze bushing sleeve bearing.

Rubber Mounts

Rubber mounts (Fig. 2–6) fit between the base of the motor and mounting rails, and reduce the noise level of the motor. Two motor-mounting rails are shown in Fig. 2–7. A large variety of sizes of rubber mounts is available to fit a number of frames. Some types of frame mounts are used to make a motor fit other forms of installation. Rubber mounts should be bought as needed.

Fiber Washers

Fiber washers are used to remove the end play from motors. They take up the space that exists due to manufacturing tolerances. In the

Fig. 2–5. Shaft adapter.

Fig. 2–6.   Rubber motor mounts.

process of disassembling a motor it is easy to misplace a fiber washer, so it is wise to have a ready supply in stock. They are sold to repair shops in kits containing 48 different sizes.

Rubber Grommets

Rubber grommets prevent the electrical cord from becoming shorted by contacting the frame of the motor. They fit into the metal case and insulate the hole by providing a rubber covering. A ⅜-inch and a ½-inch bore are usually sufficient for most uses. The ⅜-inch bore needs a ½-inch hole and the ½-inch bore grommet requires a ¾-inch hole. This is necessary because the outside diameter is ⅞ inch. A good supply of rubber grommets should be kept on hand.

Replacement Switches

In split-phase, capacitor-start, and permanent-split capacitor motors, centrifugal switches are needed to remove the start winding from the circuit after the motor has reached run speed. These switches can fail for a number of reasons. Wear, too much oil, dirt, pitted contacts, and overload on the motor can all cause failure.

Centrifugal switches have various shapes, as shown in Fig. 2–8. The ten switches in the figure are made for Westinghouse motors. One style of centrifugal switch with two operative parts is shown in close-up in Fig. 2–9. In Fig. 2–10 the switch is shown in both the start and

Fig. 2–7.   Motor-mounting rails.

REPLACEMENT SWITCHES

Factory Parts for Split Phase and Capacitor Motors, Nema 48 and 56 Frames

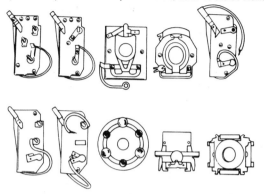

Fig. 2–8. Replacement switches for Westinghouse motors.

running positions. The contacts are closed when the motor is not operating. When the motor is started, the movement of the rotor (centrifugal force) causes the round slider to move inward on the shaft. As the slider moves inward, it causes the switch contacts to open and the start winding circuit is broken. If the motor is turned off, the slider will move out again. It is forced by spring action to close the switch contacts of the start winding. The motor is then ready to start again with the start winding in the circuit.

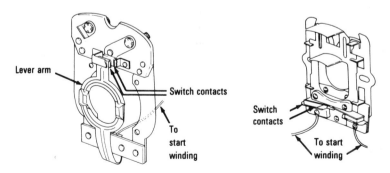

Fig. 2–9. One type of centrifugal switch (both sides).

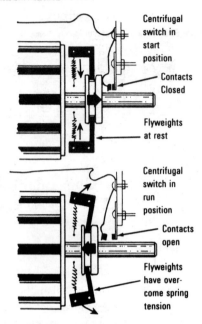

Fig. 2–10. Operation of the centrifugal switch.

If the centrifugal switch does not close when the motor is stopped, the rotor will not move when the power is turned on. The motor will hum but will not actually start. However, if you give the motor shaft a start by hand—moving it either clockwise or counterclockwise—the motor will start.

Electrolytic Capacitors

The capacitor-start motor is basically a split-phase motor with two separate windings in parallel connection: the main winding and the start winding. In the capacitor-start motor, an electrolytic capacitor is inserted in series, with the start winding in the start position, to increase the starting torque and reduce the starting current. When replacing the capacitor in the motor, use only the electrolytic capacitor marked for motor use.

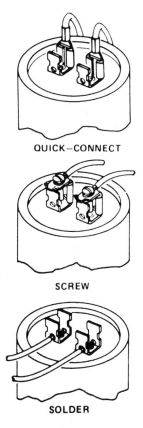

QUICK−CONNECT

SCREW

SOLDER

Fig. 2–11. Three types of connections to an electrolytic capacitor used in a motor circuit.

There are two types of electrolytic capacitors—polarized and non-polarized. The polarized type is used in electronics. It has a positive (+) symbol on one lead and negative (−) symbol on the other. It may also have a red dot or some other type of code for the positive lead. The polarized type will operate on DC only, and should not be used with AC at all.

A nonpolarized electrolytic capacitor can be made by placing two polarized types in series. However, they must be set up in series oppo-

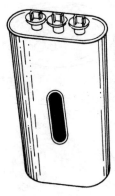

Fig. 2–12. Oil-filled capacitor for use with motors.

sition; in other words, connect them so that positive (+) goes to positive (+) and the two negative (−) terminals go to the AC circuit.

One of the problems with a *substitute* electrolytic capacitor is making sure that the capacitance is correct. If the two capacitors are connected in series, the capacitance is reduced. For instance, if two 100
μF capacitors were connected in series, they would produce a total
capacitance of only 50 μF. However, the working voltage DC is increased. Two capacitors each with 50 WVDC and placed in series will
combine to make 100 WVDC capability.

Capacitor-start (CS) motors use only one electrolytic to start. The
capacitor is usually round in shape and has the size indicated in μF.
The voltage is marked in white letters and numbers if the case is made
of black Bakelite®. Fig. 2–11 shows three types of connectors for an
electrolytic capacitor used on electric motors.

The permanent-split capacitor (PSC) motor may use one oil-type

Fig. 2–13. Capacitor located on top of a motor. *(Courtesy Leeson)*

capacitor (usually an oblong type) for both starting and running (Fig. 2–12). Some motors require two capacitors to start and one capacitor to run. These motors use one oil-type and one dry electrolytic-type capacitor in the start position. Only the oil type is needed while the motor is running.

There are two types of electrolytics—dry and wet. The dry type uses a paste material and is not truly dry. The wet type uses oil as the electrolyte.

The capacitor is usually located on top of the motor and has a metal cover plate or enclosure. For refrigerators, air conditioners (large units), and washing machines, the capacitor may be mounted somewhere other than on the motor itself (Fig. 2–13).

Insulating Enamel

Insulating enamel can be used as a protective coating for frames and end shields and as a gasket cement and sealant for oil, water, and connections. The enamel can also be used on windings, coils, commutator ends, and bus bars. It produces a tough, flexible, oilproof, and waterproof film. Insulating enamel comes in spray cans and is quick drying.

Cord

A supply of heavy-duty cord should be on hand to replace any that may be dried out and cracked. Three-conductor cord will have white,

Table 2–1. Cord Sizes and Current-Carrying Capabilities

Cord Sizes and Uses			
	Type	Wire Size	Use
Ordinary lamp Cord	POSJ SPT	No. 16 or 18	In residences for lamps or small appliances.
Heavy-duty—with thicker covering	S or SJ	No. 10, 12, 14, or 16	In shops, and outdoors for larger motors, lawn mowers, outdoor lighting, etc.

Ability of Cord to Carry Current (2- or 3-wire cord)			
Wire Size	Type	Normal Load	Capacity Load
No. 18	S, SJ, or POSJ	5.0 amp. (600W)	7 amp. (840W)
No. 16	S, SJ, or POSJ	8.3 amp. (1000W)	10 amp. (1200W)
No. 14	S	12.5 amp. (1500W)	15 amp. (1800W)
No. 12	S	16.6 amp. (1900W)	20 amp. (2400W)

Selecting the Length of Wire		
Light Load (to 7 amp.)	Medium Load (7–10 amp.)	Heavy Load (10–15 amp.)
To 15 ft.—Use No. 18 To 25 ft.—Use No. 16 To 35 ft.—Use No. 14	To 15 ft.—Use No. 16 To 25 ft.—Use No. 14	To 15 ft.—Use No. 14 To 25 ft.—Use No. 12 To 45 ft.—Use No. 10

Note: As a safety precaution, be sure to use only cords which are listed by Underwriters Laboratories. Look for the UL seal.

Table 2–2. Individual Branch Circuit Wiring for Single-Phase Induction Motors

Motor Data		Copper Wire Size (Minimum AWG No.)				
		Branch Circuit Length				
H.p.	Volts	0–25 feet	50 feet	100 feet	150 feet	200 feet
1/6	115	14	14	14	12	10
	230	14	14	14	14	14
1/4	115	14	14	12	10	8
	230	14	14	14	14	14
1/3	115	14	12	10	8	6
	230	14	14	14	14	12
1/2	115	14	12	10	8	6
	230	14	14	14	14	12
1/4	115	12	10	8	6	4
	230	14	14	14	12	10
1 1/2	115	12	10	8	6	4
	230	14	14	14	14	10
1	115	10	10	6	4	4
	230	14	14	12	10	8
2	115	10	8	6	4	
	230	14	12	12	10	8
3	115	6	6	4		
	230	10	10	10	8	8
5	230	8	8	8	6	4

black, and green wire. The green wire is the safety wire and should be connected to the frame of the motor. The rating of the cord needs to be checked. Then, it is important to know the current rating of the motor to see if the motor and cord are matched properly.

Every cord has a maximum allowable current-carrying capacity (Table 2–1). Cords for heavy motors and high-wattage appliances must be heavy enough to carry the needed current. Overloaded cords overheat, wasting power and often causing the motor to run at lower than normal speeds. Of course, there is always the possibility of fire from an overheated cord.

Table 2–3. Individual Branch Circuit Wiring for Three-Phase Squirrel-Cage Induction Motors

Motor Data		Copper Wire Size (Minimum AWG No.)				
		Branch Circuit Length				
H.p.	Volts	0–25 feet	50 feet	100 feet	150 feet	200 feet
½	230	14	14	14	14	14
	460	14	14	14	14	14
¾	230	14	14	14	14	14
	460	14	14	14	14	14
1	230	14	14	14	14	14
	460	14	14	14	14	14
1½	230	14	14	14	14	12
	460	14	14	14	14	14
2	230	14	14	14	12	12
	460	14	14	14	14	14
3	230	14	14	12	10	10
	460	14	14	14	14	14
5	230	12	12	10	10	8
	460	14	14	14	14	14
7½	230	10	10	10	8	6
	460	14	14	14	14	12
10	230	8	8	8	6	4
	460	12	12	12	12	12
15	230	6	6	6	4	4
	460	10	10	10	10	10
20	230	4	4	4	4	3
	460	8	8	8	8	8
25	230	4	4	4	3	2
	460	8	8	8	8	8
30	230	3	3	3	2	1
	460	6	6	6	6	6
40	230	1	1	1	1	0
	460	6	6	6	6	6
50	230	00	00	00	00	00
	460	4	4	4	4	4

Motor Wiring for Installation

All wiring and electrical connections should comply with the *National Electrical Code* (*NEC*) and with local codes and sound practices.

Use of undersized wire between the motor and the power source will adversely limit the starting and load-carrying abilities of a motor. Recommended mininum wire sizes for motor branch circuits are given in Tables 2–2 and 2–3.

PART TWO

Fractional-Horsepower Motors (Basic Types)

CHAPTER 3

Small DC and Universal Motors

DC motors are made in series, shunt, and compound configurations. The series motor is also used on AC. When used on AC or made to be able to use both AC and DC, it is called a universal motor.

The latest development in DC motors is the PM, or permanent magnet, type. Instead of a field coil, this motor uses a permanent magnet and a wound armature.

Shunt DC Motors

The shunt DC motor (Fig. 3–1) is one of the most versatile of DC types. It is a relatively constant-speed motor. Solid-state circuitry can be used to control its speed over a wide range. It has a wound field coil, or coils, and a wound rotor (armature). It does, however, use brushes and a commutator. The armature is connected across (in shunt or in parallel with) the field windings.

These motors are available in base speeds of 1140, 1725, 2500, and 3450 rpm. They can be wound for almost any speed or for special applications.

The shunt motor is reversible at rest or during operation. Reversing is done by simply reversing the polarity of the armature or the field

Fig. 3–1. Shunt wound DC motor. Note the snap-together case. *(Courtesy Voorias)*

voltage. Usually it is the armature that is reversed, since the field winding has high inductance. High inductance can cause excessive arcing at the switch contacts as the voltage is reversed.

The shunt DC motor is used for many purposes. It can be used to run windshield wipers on cars, the fans in car heaters, and special types of printing equipment. Printing presses with DC power available have better control of the printing process with DC shunt motors. They are almost constant speed when the voltage and load are kept constant.

Plugging (reversing the motor when it is running) can be harmful to the motor. This subjects the armature to approximately twice the rated voltage. Dynamic braking (placing a short across the armature when the power is removed) should be handled with caution.

Brush life is good on a shunt motor. This and the armature commutator segments are the main concern for maintenance. Plugging and dynamic braking can severely limit the life of the brushes. Keep in mind that one brush will wear faster than the other. This is normal operation for a shunt DC motor (Fig. 3–2).

Compound Motors

Since the DC compound motor is usually larger than one horsepower, it will be considered just briefly here. The compound motor has a series and a shunt coil and can be connected in various combinations. The SCR speed control and the permanent-magnet motor have replaced the compound motor in fractional-horsepower sizes.

The compound motor has constant speed and high starting torque. It has the qualities of both series and shunt motors and can be used for such things as elevators, grain mill operations, and anywhere there is a need for good starting torque and fairly constant speed.

Compound motors are designed in two configurations, cumulative compound and differential compound. Cumulative compound DC motors have the two fields wound in the same direction. Speed depends on the sum of the two fields. Torque is higher than that for a shunt motor, but speed regulation is less than that for a shunt motor.

Differential compound DC motors have the coils wound in the opposite direction; that is, the shunt coil is wound in one direction and the series coil is wound in another. Speed depends on the difference of the two fields. Torque is lower than that for a shunt motor for the same amount of armature current. As a load is applied, the speed *increases*.

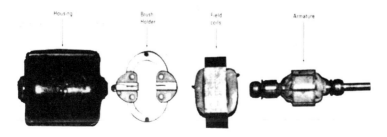

Fig. 3–2. Notice the brushholder of the small shunt DC motor.

Occasionally the differential compound motor has been specially made to perform at a particular point. The term *flat-compounded* is used in this case. It means that a relatively flat speed-torque curve results from the design.

Permanent-Magnet Motors

The permanent-magnet (PM) motor has a permanent magnet that replaces the field coil. That means this type of DC motor will have only a wound armature (Fig. 3–3). The magnet is interesting since it takes advantage of the latest technology in oriented strontium ferrites (ce-

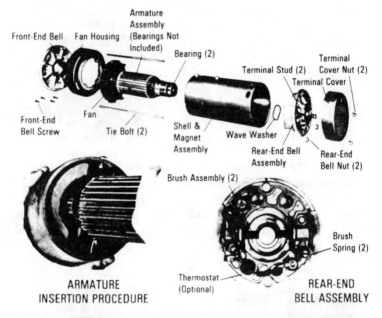

Fig. 3–3. Exploded view of a permanent-magnet motor. When disassembling for maintenance, use a clean bench free of steep parts or chips. When replacing brushes, check the commutator for wear. If commutator is worn down more than $\frac{1}{32}$ inch on the diameter, turning and undercutting is recommended. Usually three sets of brushes can be used for one commutator turning. *(Courtesy Doerr)*

ramic magnet). In some cases, rare earth metals are used for the magnet, but this results in a rather expensive type of motor.

The best reason for using a PM motor is its size. It is physically smaller than a comparable shunt-type DC motor (Fig. 3–4). This type of motor produces relatively high torques at low speeds. The PM motors are often used as substitutes for gearmotors. Permanent-magnet motors cannot be continuously operated at the high torques they are able to generate because they will seriously overheat.

Permanent-magnet motors draw less current from the battery than the shunt type and are therefore more efficient. The permanent magnet also produces some braking. The coasting of the armature after power is removed can be stopped by using dynamic braking. This is done simply by shorting across the armature. This type of motor can be reversed simply by reversing the polarity of the armature leads.

There are some limits of this type of motor. It can lose some of its magnetism at temperatures below zero, and there is some change in the working flux when exposed to high temperatures. However, there are many applications for PM motors. They are used in the automobile industry for window lifts, heaters, blowers, defrosters, seat adjusters, windshield wipers, and rear deck defrosters. In marine equipment, you will find them used in variable-speed pumps, water pumps, fishing

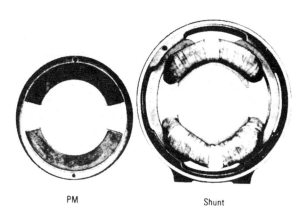

PM   Shunt

Fig. 3–4. Note the size difference, especially the outside diameter, of the PM motor as compared to the shunt wound motor with the same diameter for the armature. *(Courtesy Bodine)*

reels, trollers, winches, and blowers. Other applications include office machines, door-latch mechanisms, fans, motorized valves, and cordless appliances. Armatures can be wound to handle from 60 to 230 volts. About the only maintenance is the brushes. The commutator may need some attention after prolonged use. Remember, these are generally designed for intermittent duty. The bearings in most are bronze with a felt reservoir for holding lubrication. Some are available with sealed ball bearings.

Series or Universal Motors

As already stated, the series DC motor is used as a universal motor since it can be designed to run on either AC or DC. The series wound motor is among the most popular of fractional and subfractional motors. It can deliver high speed and high torque at starting, and provides a wide variety of speeds when used with speed controllers. At first glance, the series motor may look just like the shunt type of DC motor. However, upon looking closely, you can see that the windings are different. The difference between operation on 50 Hz as compared to 60 Hz is negligible.

All series motors are not necessarily universal. Some may have been designed for operation on DC only; they may not perform well on AC or may fail completely. Check the nameplate to be sure. Usually the universal motor will run slower on AC than on DC. In some cases it is necessary to place a resistor in series with the power source to slow down the DC-operated motor to where it would operate normally on AC. This is the case only when speed is critical. Placing a resistor in series with a universal motor does dramatically reduce its starting torque. A good example of this is the sewing machine. When the foot-pedal speed control is used (it is nothing more than a resistor being varied by the foot pedal), its resistance causes the sewing machine to have little or no torque. Most sewing machines have a hand-operated wheel so the motor can be started manually.

Series motors are the least expensive in terms of dollar per horsepower of any type of motor, which is one reason why you find them in almost all household appliances. They are the only small motors capable of doing more than 3600 rpm. In fact, they will operate at 10,000 to

20,000 rpm. *Caution: They should not be operated without a load.* Once they are loaded, you can hear the speed decrease.

Direction of Rotation

One thing you should keep in mind when working with universal motors: the direction of rotation is important. Universal motors are usually supplied with one direction specified. This is primarily to improve operating efficiency, since the brushes seat at a point where the armature is rotating in a given direction. If the direction is changed, the nature of the wear on the brush will likewise change. Therefore, the brush life is affected by the direction of rotation.

Some Disadvantages

The series motor does not have good speed regulation. It will vary in speed with the load applied. Bearings and brushes wear faster when the armature turns fast. Most home appliance motors have from 200 to 1200 hours of operation on brushes before they have to be replaced. Brush replacement is the most common maintenance procedure needed with this type of motor. Compare the characteristics of the series DC motor with those of other motors (Table 3–1).

Table 3–1. Motor Characteristics

Type	Shunt Wound	Series Wound	Compound Wound
Duty	Continuous	Intermittent	Continuous
Power Supply	DC	DC	DC
Reversibility	At rest or during rotation	Usually unidirectional	At rest or during rotation
Speed	Relatively constant and adjustable	Varying with load	Relatively constant and adjustable
Starting Torque	125 to 200% of rated torque	175% and up of rated torque	125% and up of rated torque
Starting Current	Normal	High	Normal

Types of Universal Motors

Series motors used for AC and DC are made in a number of different configurations. Fig. 3–5 shows the simplest of the universal motors. It has a C-frame type of construction.

Fig. 3–6 shows an example of the A-frame series of universal motor. It has two coils connected in series with each other and also in series with the armature through brushes.

Fig. 3–7 shows the E-frame universal motor. Check its characteristics in Table 3–2. This one does not have a fan mounted on the shaft; it may be totally enclosed without a fan. Note the shape of the armature.

In Fig. 3–8 you will find the oversized B-frame type of universal motor with a fan. Note the number of leads from the coil. This also has a brushholder mounted on the frame. The PQ-frame motor is shown in Fig. 3–9, while the K-frame universal motor is shown in Fig. 3–10. Note, in Table 3–2, that the C-16 on the C-frame type of motor means it has 16 segments to the commutator; the C-20 has 20 segments; etc. The number 224-18 for the A-frame means 2.24 inches in diameter for

Fig. 3–5. Universal motor. *(Courtesy Voorlas)*

Fig. 3–6. A-frame universal motor. *(Courtesy Voorlas)*

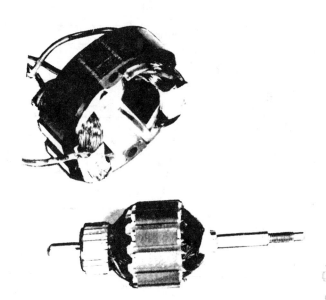

Fig. 3–7. E-frame universal motor. *(Courtesy Voorlas)*

Table 3–2. Series Motor Types and Characteristics

C-Frame	Max. Watt Output	RPM	Normal Watt Output	RPM
C-16	50	10000	30	20000
C-20	65	9000	40	17500
C-24	85	10000	55	15000
A-Frame				
224-18	125	9500	90	14000
224-24	140	8500	100	13000
224-28	165	8000	110	12500
224-32	230	8000	125	14000
B-Frame				
243-16	130	8000	105	12500
243-20	180	7500	135	11500
243-24	220	8000	160	12500
243-28	260	7500	190	11500
262-16	130	8000	105	12500
262-20	180	7500	135	11500
262-24	220	8000	160	12500
262-28	260	7500	190	11500
E-Frame				
287-16	180	7000	120	13000
287-20	250	8000	150	13000
287-24	285	7500	175	12500
287-28	320	6000	200	11000
287-30	335	6000	215	10000
287-34	370	6000	240	10000
K-Frame				
318-14	350	6000	170	12000
318-17	460	6000	220	12000
318-19	570	6000	280	12000
318-22	740	6000	340	12000
318-33	880	6000	430	12000

Table 3–2.   Series Motor Types and Characteristics (cont.)

Oversize B-Frame	Max. Watt Output	RPM	Normal Watt Output	RPM
P.Q.-Frame				
368-16	330	7500	200	13000
368-20	450	6000	230	11000
368-24	550	6000	280	11000
368-28	675	5500	325	11000
368-32	800	5500	400	10000
368-36	1000	4100	500	12000
368-40	1100	4100	700	9500

(Courtesy Voorlas)

Fig. 3–8.   B-frame universal motor. *(Courtesy Voorlas)*

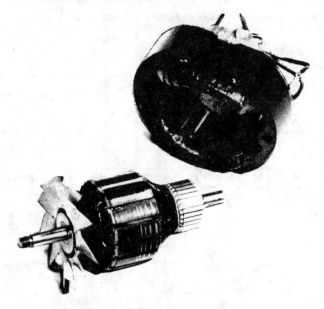

Fig. 3–9. PQ-frame universal motor. *(Courtesy Voorlas)*

the outside measurement and 18 segments in the commutator. Follow the same interpretation for corresponding numbers in the rest of the table. Note the variation in rpm as the maximum and normal wattages are compared. The larger the load, the slower the motor. This, again, is typical of the series motor.

Armatures

In order to keep down vibration and noise, it is necessary to balance the armature since it rotates at about 20,000 rpm. Fig. 3–11 shows how a typical armature (on the left) has been balanced by drilling holes in the laminated core. The number of holes and the depth determine the balance. The fan blades are press-fitted onto the shaft. Note the white silicon used to hold the wires from the coils in place near the connection to the commutator segments. The centrifugal force is so great at 20,000 rpm that it can cause the windings to separate, since the force of the rotating armature will be exerted on anything not held to

Fig. 3–10. K-frame universal motor. *(Courtesy Voorlas)*

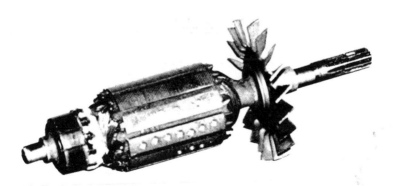

Fig. 3–11. Armature for a hand-held electric drill. Note the holes drilled in the laminations for purposes of balance.

the shaft. If the connections to the commutator are broken, the circuit is broken at this particular coil and will cause some rough operation and excessive arcing. The armature on the right uses the conventional method of connection—more insulation (varnish). The silicon is supposed to be an improved method. Most connections are now made to the commutator segments by welding and not soldering.

The Hand Drill

One of the most popular tools around the house is the hand drill (Fig. 3–12). It is a prime example of the use of a universal motor. This series motor is loaded by the gear reduction located near the Jacob's chuck—or the end where the drill bit is attached to the tool. Note how the armature is inserted into the coil mounted on the drill housing (Fig. 3–13).

Most of the maintenance of this machine is caused by the misuse of the tool. If the brushes are sparking too much, it is time to remove them and the armature, smooth the commutator segments with sandpaper (*not* emery paper) and blow them clean, then replace the brushes. The grease in the gear box is usually sufficient for the life of the drill. It can be damaged by extremes in heat, however, and may need to be replaced.

Bearing

Fig. 3–12. Disassembled hand drill. Note that the bearing is the rounded piece of bronze near the gasket.

Fig. 3–13. Coils of the series motor used on a hand drill. Farther back you can see the brushholder.

Speed Control

Most hand-held electric drills are variable-speed. The speed control unit is nothing more than a variable resistor being controlled as the switch is pulled back toward the handle (Fig. 3–14). The gear-operated resistor changes the electronic circuit so that the SCR operates properly and increases the speed as you pull harder on the trigger. The SCR controls the amount of current allowed through the coils. A heat sink is used to dissipate the heat for cooler operation of the SCR. Keep in mind that the SCR rectifies the AC and produces DC; this means that the drill is now operating on DC. In most instances this also means a greater efficiency since the series motor is more efficient on DC.

Fig. 3–14. Variable-speed control for a series motor used as a hand drill. *(Courtesy Black & Decker)*

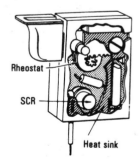

Slower speeds are used for starting holes without skipping or center punching. A slower speed can also be used for mixing paints and drilling ceramic tile. Medium speed is used for drilling plastics and metals. Faster speeds are used for drilling wood and for powering accessories such as buffers and grinders. Once the trigger has been squeezed all the way back to the handle, it clicks a switch and takes the SCR out of the circuit and the drill operates on the full 120 volts.

Portable Saws

Portable saws are powered by a universal motor. The arrow in Fig. 3–15 shows where to look on the portable saw to check for excess sparking at the brushes. Too much arcing will cause serious damage to

Fig. 3–15. The arrow points to the area to check for excess arcing of the brushes.

the commutator. If the motor is stalled, it is subject to overheating and excess current flow through the windings of the armature and the coils.

Proper maintenance requires keeping the ventilation area free from sawdust and other types of dirt that may cause damage to the bearings and get caught between the armature and the stationary part of the motor.

Repairing a Series (Universal) Motor

Upright vacuum cleaners have been in use for many years. Many of them are out of service simply because the motor was not properly maintained. In this section we are going to disassemble the motor of an upright vacuum cleaner to point out some of the problems related to the operation of such a machine.

The upright we are going to disassemble is shown in Figs. 3–16 and 3–17. The motor is located under the hood, which means that the hood must be removed before the motor can be reached for mainte-

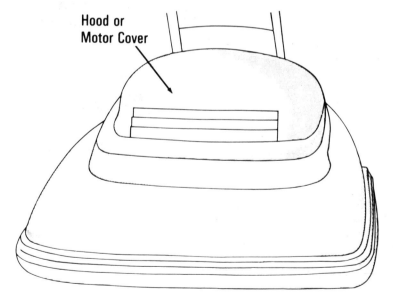

Hood or Motor Cover

Fig. 3–16. Upright vacuum cleaner using a universal motor.

VACUUM CLEANER

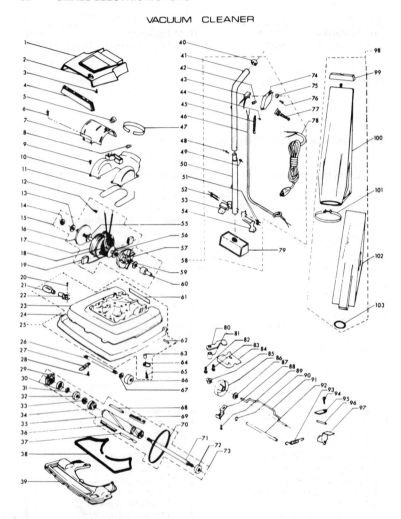

Fig. 3–17. Exploded view of vacuum cleaner.

nance or repair. Note the location of the brushes. They are easily located by just following the two wires that come from the motor housing. Fig. 3–18 shows a screwdriver being inserted into the screw that holds the brush cover in place. Once the brush cover is removed, a

Upright Vacuum Cleaner

Key	Description	Key	Description
1	Top Cover w/#2	52	Screw (2)
2	Trade Label ("Flor Genie")	53	Bifurcated Handle Support (Right)
3	Screw (#8-18 ω $3/4$" Lg. Rd Hd Thd	54	Bifurcated Handle Support (Left)
	Form, C.R.) (3)	55	Shoulder Washer, Fan (R.H.)
4	Lens	56	Shoulder Washer, Fan (L.H.)
5	Bushing	57	Suction Fan (Left Side) w/#56, #59
6	Screw (#10-16 ω $5/8$" Lg.)(3)	58	Handle Assembly Complete
7	Motor Hood Complete	59	Washer - Fan (L.H.)
8	Solderless Connector (2)	60	Motor Shaft Pulley
9	Screw (#10-16 ω $5/8$" Lg.)(11)	61	Nozzle Adjustment Decal
10	Blower Cover w/#11	62	Rivet Bumper (6)
11	Gasket	63	Spacer (3)
12	Solderless Connector (3)	64	Retainer Cup - Access Plate (3)
13	Washer - Fan (R.H.)	65	Screw (#10-16 ω $5/8$" Lg.)(3)
14	Nut, Suction Fan	66	Wheel - Front (2)
15	Motor Complete, Universal	67	Washer (4)
16	Suction Fan (R.H.) w/#13, #55	68	Beater Bar (Right Side)
17	Field Core w/#19	69	Bristle Assembly (Belt Side)
18	Armature w/Bearings (Threaded Ends)	70	Belt
19	Carbon Brush Assembly (Set of 2)	71	Shaft, Floor Brush
20	Rivet - Light Socket	72	Cup (Belt Side) - (Marked w/"B")
21	Light Bulb	73	Floor Brush Complete (Universal)
22	Light Socket	74	Switch Box
23	Chassis	75	Switch Lock Nut
24	Bumper	76	Screw (Switch Box Mounting)
25	Chassis Assembly Complete	77	Cord Strain Relief
26	Axle - Front Wheel	78	Cord
27	Retaining Plate - Front Axle (2)	79	Handle Support Top
28	Screw (#10-16 ω $7/16$" Lg.)(2)	80	Elevating Cam Lever
29	End Cap (2)	81	Spacer
30	Cup (L.H. Threads)	82	Screw (#10-16 ω $5/8$" Lg.)
31	Felt Washer (2)	83	Screw (#10-16 ω $7/16$" Lg.)
32	Ball Beaing (2)	84	Wheel & Cam Retaining Plate
33	Loading Spring (2)	85	Wheel Cap (2)
34	Bearing Retainer (2)	86	Rear Wheel (2)
35	Brush Holder	87	Washer (2)
36	Bristle Assembly (Right Side)	88	Retaining Plate - Rear
37	Beater Bar (Belt Side)	89	Screw (2)
38	Gasket Access Plate	90	Retaining Ring
39	Access Plate Assembly w/#38	91	Shaft - Rear Wheel
40	Plug Button	92	Trunion Arm Shaft
41	Tubular Handle (Upper)	93	Tension Spring - Pedal
42	Solderless Connector (3)	94	Screw
43	Switch	95	Retainer - Pedal Shaft
44	Screw (Tension Spring) (#6-20 ω $1/4$" Lg.)	96	Shaft - Pedal
45	Tension Spring	97	Pedal - Handle Release
46	2-Conductor Cord	98	Dust Bag Assembly
47	Gasket, Bag Assembly	99	Bag Top
48	Screw (Upper Handle)	100	Dust Bag Assembly w/#47
49	"T" Nut (Curved Flange)	101	Dust Bag Strap Assembly
50	Tubular Handle (Lower)	102	Disposable Inner Bag
51	"T" Nut (Curved Flange)(2)	103	Garter Spring

Fig. 3–18. Screwdriver is inserted into the screw head that holds the brush cover in place.

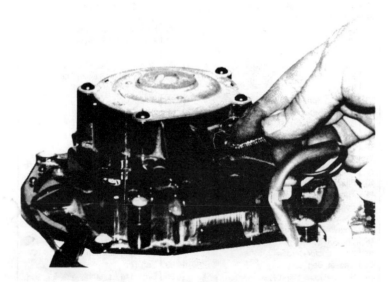

Fig. 3–19. Removing the brush. Note that it is spring-loaded.

small piece of brass holds the spring-loaded brush in its channel. Remove the brush (Fig. 3–19). Move over to the other side of the motor and remove the other brush.

Next, remove the top plate from the motor. Remove the four screws. While you have the cover plate off, place a drop of machine oil in the bearing (Fig. 3–20). You must hold the other end of the motor armature if you want to remove the small fan on top of the armature (Fig. 3–21).

In order to get at the armature from the bottom of the sweeper it is necessary to remove the cover plate shown in Fig. 3–22 by sliding the two handles shown here by arrows. The ends are hooked under the spring at the back of the sweeper, and the whole plate comes off easily. Once it is off, you can see the slot in the motor armature. In Fig. 3–23, you can see that the belt also needs replacing. Some fuzz from carpeting is also evident around parts of the sweeper. This is a good time to clean up the surface areas.

Note, in Fig. 3–24, that the motor used here has a C-frame. Only

Fig. 3–20. Oiling the top bearing on the motor.

Fig. 3–21. Unscrew the top fan from the armature to be able to clean the armature. You have to hold the bottom of the armature with a screwdriver while doing this.

Fig. 3–22. In order to get at the bottom of the armature you have to remove this cover plate. Pull outward on the two tabs shown by arrows.

Fig. 3–23. The fan or blower for the vacuum is shown. Note the condition of the brush drive belt.

one coil is used to furnish the power for the machine. The armature has slots in the commutator that need attention. Clean the armature and reassemble in reverse order. Make sure the brushes are reinserted or new ones are seated properly. Clean the area around the armature. Make sure the entire area around the motor is clean before you reas-

Fig. 3–24. **The armature is now visible can can be cleaned or checked for wear.** *Note:* **One coil is all that is used for this type of vacuum cleaner.**

semble. Brushes will need attention in 500 to 2000 hours of operation. The heavier the carpet and the more frequently the vacuum is used, the more attention it will need.

Tank Vacuum Cleaners

The tank-type vacuum cleaner has a slightly different motor design. Tank-type motors are made in a number of designs for the household cleaner. Replacement brushes are available at electrical supply stores that sell motor parts. Commercial and industrial vacuum cleaners use a slightly different type of motor housing, but all use universal motors. They are just different in the shape and design needed for a specific purpose.

Motor Installation and Mounting

There are many different methods of mounting vacuum cleaner motors. The ideal method involves clamping the fan case between two sponge rubber gaskets (Fig. 3–25). The gaskets should be compressed

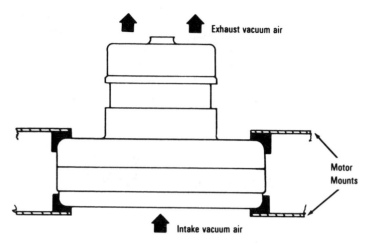

Exhaust vacuum air

Motor
Mounts

Intake vacuum air

Fig. 3–25. Motor installation and mounting for a vacuum. *(Courtesy Northland)*

to ensure an airtight seal for the vacuum chamber. This method also acts as a shock mounting for the motor.

In clamping the fan case between the two mounting gaskets, care should be taken to allow sufficient pressure to resist start-up torque of the motor. Too much compression of the gaskets will decrease the shock protection provided by the sponge rubber gaskets.

Another method of mounting uses only one gasket between the motor fan case and the mounting flange (Fig. 3–26). Screws are assembled through the mounting flange into the fan case, compressing the gasket and providing a vacuum seal. This method does not provide as good a shock mounting as the preceding method. Because of the vacuum created by the motor opposing the seal, the clamping must be tighter in order to ensure a good seal.

Bypass vacuum motors should be mounted in a manner which keeps the working air separated from the cooling air in order to prevent overheating of the motor due to recirculation. Fig. 3–27 outlines the preferred method of mounting, which separates the intake and exhaust of both the working and cooling air systems. Caution should be taken in mounting so that cooling air is not restricted from flowing around the upper bearing, since this may cause overheating and severely reduce bearing life.

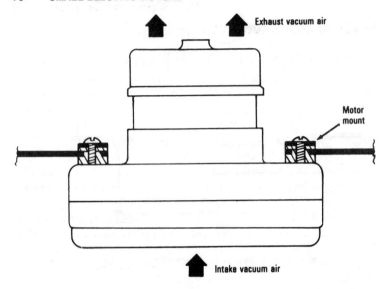

Fig. 3–26. **Another method of mounting a vacuum motor.** *(Courtesy Northland)*

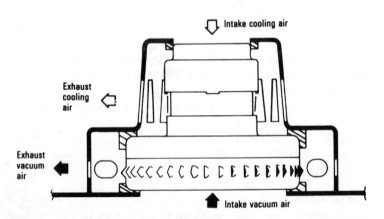

Fig. 3–27. **Operation of a properly-mounted vacuum motor.** *(Courtesy Northland)*

Miscellaneous Information

It may be helpful, in some instances, to know that vacuum motors may be combined to achieve additional performance. Two or more motors may be connected in (air) series. This arrangement significantly increases vacuum performance with little increase in the cubic feet per minute (CFM) of air moved. Connecting two motors in (air) series is normally accomplished by mounting two or more motors in a long tube. The lower motor can be of the flow-through type and the upper units should be of the bypass type. This arrangement of a flow-through and a bypass prevents overheating of the upper unit due to the passage of heat from the lower unit.

Vacuum motors may also be connected in (air) parallel. This arrangement significantly increases CFM with little increase in vacuum. Connecting motors in (air) parallel is accomplished by mounting the motors on the cover of a large container, allowing them to work in the same airspace.

When lower performance is wanted than is available in a given vacuum motor, it can operate on a lower voltage. Running on a lower voltage will also increase the life expectancy of the motor.

Reducing the speed of a motor will also increase the life expectancy. A rough rule of thumb is that when the speed is reduced by 50 percent, the brush life is tripled.

CHAPTER 4

Split-Phase Motors

One form of the fractional-horsepower motor is the split-phase induction motor. This motor has two sets of windings: the *run* (main) winding and the *start* (auxiliary) winding. The run winding is the main workhorse of this motor. The start winding is used only when the motor is started; therefore it is called an auxiliary winding. For a two-pole motor, the start winding is placed 90° electrically from the main winding (Fig. 4–1). For a four-pole vector, the angle would be 45°. It is connected in parallel to the run winding (Fig. 4–2). Without the auxiliary start winding, this motor would have no starting torque. However, with the start winding, the rotor will reach between 67 to 75 percent of the top or synchronous speed. At this speed, the motor develops a good running torque and the start winding is no longer needed. The start winding is disconnected from the motor circuit by an automatic starting switch within the motor case. This is usually a centrifugal switch.

Starting a Single-Phase Motor

In a single-phase AC motor (Table 4–1), the field pulsates instead of rotating as does a two- or three-phase motor. No rotation of the rotor

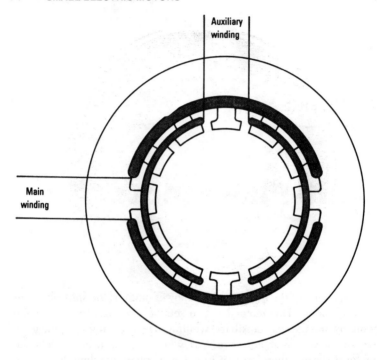

Fig. 4–1. Two-pole, single-phase stator with auxiliary (start) winding and main (run) windings. Note the 90° displacement of the centers of the windings.

occurs. However, a single-phase pulsating field may be seen as two rotating fields revolving at the same speed but in opposite directions. The rotor will revolve in either of these directions at nearly synchronous speed. But it must be given an initial push in one direction or the other. The precise speed of the initial rotational push velocity varies widely with different motors, but a velocity of more than 15 percent of the synchronous speed is usually enough to cause the rotor to accelerate to rated speed. A single-phase motor, of which the split-phase is but one type, can be made self-starting if means can be provided to give the effect of a rotating field.

Fig. 4–2.   Four-pole, single-phase stator with start and run windings mounted on a frame. *(Courtesy Bodine)*

Starting a Split-Phase Motor

The split-phase motor has a stator composed of slotted laminations that contain an auxiliary (start) winding and a run (main) winding (Fig. 4–3). The axes of these two windings are displaced by an angle of 90° electrically. The start winding has fewer turns and smaller wire than the run winding. It also has higher resistance and less reactance. The main winding occupies the lower half of the slots and the start winding the upper half. The two windings are connected in parallel across the single-phase line that supplies power to the motor.

The motor takes its name from the action of the stator during the start period. The single-phase motor is split into two windings, or phases, which are separated by 90°. These phases contain currents displaced in time phase by an angle of approximately 15°. The current in the start winding lags the line voltage by about 30°. It is less than the

Table 4-1. Typical Characteristics of AC Motors

	HP Ratings	Full Load Speeds, rpm (60 Hertz)	Starting Torque	Breakdown Torque	Starting Current	Comparative Cost (100=lowest)	Guidelines
Shaded pole	1/65 to 1/20	1650	very low	low	low	100	Low-cost motor for light duty applications. Compact, rugged, easy to maintain.
Permanent split capacitor	1/50 to 1/3	3250	low	moderate	low	140	Very compact, easy to maintain. High efficiency, high power factor. Can operate at several speeds with simple control devices.
	1/60 to 1/6	1625					
Split phase	1/40 to 1/3	3450	moderate	moderate	high	120	For constant speed operation, varying loads. Where moderate torques are desirable, may be preferable to more expensive capacitor start.
	1/50 to 1/6	1725					
Capacitor start	1/40 to 1/3	3450	high	high	moderate	150	Suitable for constant speed under varying load, high torques, high overload capacity.
	1/50 to 1/6	1725					
Polyphase	1/30 to 1/3	3450	high	high	moderate	150	Generally suited to same applications as capacitor start motors if polyphase power is available. Gets to operating speed smoothly and quickly.
	1/75 to 1/6	1725					

(Courtesy of Robbins & Myers)

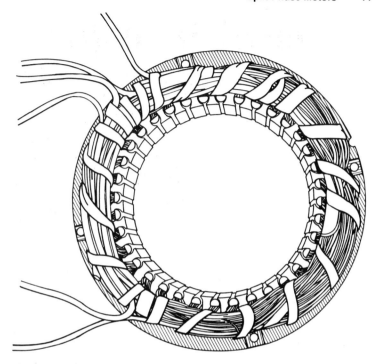

Fig. 4–3. Split-phase motor windings. Note that the start and run windings have leads that can be connected in a number of ways to affect the direction of rotation.

current in the main windings because of the higher impedance (or opposition) of the start winding. The current in the main winding lags the applied voltage by about 45° (Fig. 4–4). The total current (I_{line}) during the starting period is the vector sum of the start and main winding currents.

At the motor's start, these two windings produce a magnetic revolving field. The field rotates around the stator air gap (space between motor and stator) at synchronous speed. As the rotating field moves around the air gap, it cuts across the rotor conductors and induces a voltage. The maximum voltage is in the area of highest field intensity. Therefore, it is "in phase" with the stator field. The rotor current lags

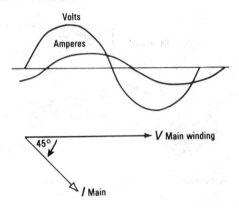

Fig. 4–4. Phase relationship between current and voltage in the split-phase motor.

the rotor voltage at the start by an angle that approaches 90° because of the high rotor reactance.

As the rotor currents and the stator field interact, they cause the rotor to accelerate in the direction in which the stator field is rotating. During acceleration, the rotor voltage, current, and reactance are reduced and the rotor currents come closer to an in-phase relationship with the stator field.

When the rotor reaches about 75 percent of synchronous speed, a centrifugally-operated switch disconnects the start winding from the line supply, and the motor continues to run on the main winding alone. Thereafter, the rotating field is maintained by the interaction of the rotor magnetomotive force and the stator magnetomotive force. These two magnetomotive forces are pictured as the vertical and horizontal vectors in Fig. 4–5C.

The stator field is assumed to be rotating at synchronous speed in a clockwise direction, and the stator currents correspond to the instant that the field is horizontal and extending from left to right across the gap. The left-hand rule for magnetic polarity of the stator indicates that the stator currents will provide a north pole on the left side of the stator and a south pole on the right side (see Fig. 4–6). The motor indicated in the figure is wound for two poles.

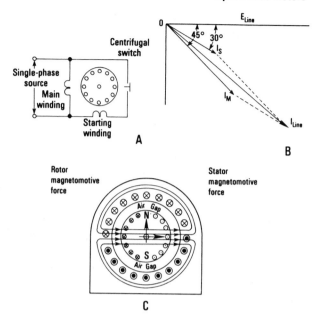

Fig. 4–5. (A) Circuit for a split-phase motor. (B) Phase relationships between current and voltage in a split-phase motor. (C) Magnetomotive force angles in a split-phase motor.

By applying the left-hand rule to find induced voltage in the rotor (the thumb points in the direction of motion of the conductor with respect to the field), we see that the induced voltage is less than the rotor voltage by an angle whose tangent is the ratio of rotor reactance to rotor resistance. This is a relatively small angle because the slip is small. When we apply the left-hand rule for magnetic polarity to the rotor winding, the vertical vector pointing upward represents the direction and magnitude of the rotor magnetomotive force. This direction indicates the tendency to establish a north pole on the upper side of the rotor and a south pole on the lower side, as indicated in Fig. 4–5C. Thus, the magnetomotive forces of the rotor and the stator are displaced in space by 90° and in time by an angle that is considerably less than 90° but sufficient to maintain the magnetic revolving field and the rotor speed.

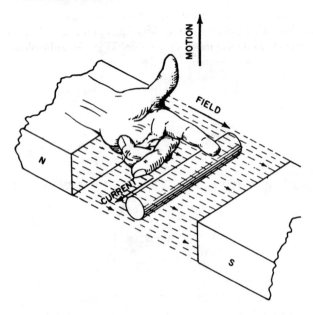

Fig. 4–6. Using the left-hand rule to explain the direction of motion of the conductor, direction of the field, and the direction of electron flow in a piece of wire under the influence of a magnetic field.

Characteristics of the Split-Phase Motor

The split-phase motor has the constant-speed, variable-torque characteristic of the shunt DC motor. Most of these motors are designed to operate on 120 or 240 volts. For the lower voltage, stator coils are divided into two equal groups which are connected in parallel. For the higher voltage, the groups are connected in series. The starting torque is 150 to 200 percent of the full-load torque and the starting current is six to eight times the full-load current. The direction of rotation of the split-phase motor can be reversed by interchanging the start winding leads. Fractional-horsepower split-phase motors are used in various machines, such as washers, oil burners, and ventilating fans.

After the start winding has been removed from the line, there is no rotating field. The rotation cannot be changed until the motor has come to rest or at least has slowed to the speed at which the automatic switch closes. Special starting switches and special reversing switches are available that can be used to shunt the open contacts of the automatic switch while the motor is running. Thus, the split-phase motor can be reversed while rotating.

Dual-Frequency Operation

The split-phase motor is not too suitable for operation on dual frequencies. In some cases the necessary compromise may be permissible. If the 60-Hz motor is designed to operate close to the flux saturation point, then the no-lead watts may be more than doubled, producing more heat, when operated on 50 Hz. An additional problem is the operating speed. This means trying to obtain the proper operating speed so that the centrifugal switch will operate and remove the start winding from the circuit. Also, finding a start relay suitable for both 50- and 60-Hz operation is difficult.

If you decide to rewind a 25-Hz motor to operate on 60-Hz current, the start-winding switch will have to be changed. The number of turns and the whole design of the motor must be considered. In most cases conversion is possible, but problems should be expected later in motor operation. Indeed, the performance of the motor may never be satisfactory.

If the change in frequency is small, say, from 50 to 60 Hz, no change in the winding of general-purpose motors is ordinarily needed unless the motor is severely overloaded. A change from 60 to 50 Hz normally requires no winding change either. If the motor has high torque, it may be necessary to increase the number of turns of the 60-Hz motor by 10 percent. This is done to obtain satisfactory operation on 50 Hz without overheating. However, when changing either a general-purpose or high-torque motor from 60 to 50 Hz, the rotating part of the starting switch should be changed. The switching torque is affected by the change in frequency. If the change is from 50 to 60 Hz and the torque requirements are not too severe, there may not be any reason for changing the windings or the switch.

The main concern is the inductive reactance (X_L) of the run winding. It changes as the frequency changes, because $X_L = 2\pi FL$. (F is for frequency, L is for inductance.) However, if the motor was designed to operate on 50 *or* 60 Hz, then the operating frequency and either the

50- or 60-Hz frequency would result in little significant heat generation that was not designed in the motor in the first place.

Motor Noise and Vibration

There are two types of motor noise: mechanical and electrical. Mechanical noise can be caused by dynamic imbalance; this means that the rotor is unbalanced. When a rotor is unbalanced, small holes are drilled into it (Fig. 4–7). The loss of the metal removed from these holes helps to balance the rotor. To check for rotor imbalance, turn off the motor and listen for a vibrating noise as the motor coasts. If there is vibration, the dynamic balance of the rotor is probably causing problems. In a split-phase motor, the centrifugal switch can also cause problems and should be checked to see if it is operating properly (Fig. 4–8).

Noise may also be created by bearings. Usually, the trouble is

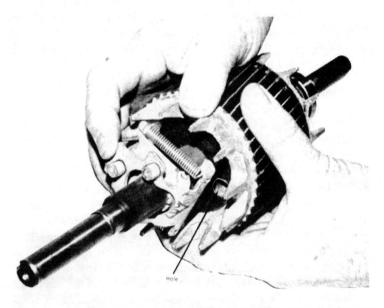

Hole

Fig. 4–7. Note the hole behind the centrifugal switch mechanism on this rotor. It is put there to aid the balance of the rotor.

Fig. 4–8. Start or centrifugal switch for a split-phase motor.

closely tied to bearing preload. Preload is the axial thrust on the outer bearing race to eliminate rattling of unloaded balls. The amount of preload that causes rattling noises is very low—below 2 lb. for most motors. Noise levels for ball-bearing motors will not fall below 40 dB in any event. Also, a slight variation in noise level between identical motors is normal; thus, do not expect to find a very quiet ball-bearing motor (Fig. 4–9).

Sleeve bearings produce much lower noise levels than do ball bearings. If noise must be kept low, sleeve bearings can be used in place of ball bearings. The biggest problem with sleeve bearings is to control noise from the thrust washers. These washers have a tendency to produce an intermittent scraping sound. Another problem is that bearing knocking or pounding occurs after the motor has been operated for a long time. This happens when the oil for the bearings has been thinned by heat.

Fans can be a major source of noise. Even fans with low-speed motors can produce noticeable levels of noise. The swish or rumble from an air exhaust opening can be annoying to some people. In electric motors, fans are used to keep air flowing over the rotor and stator windings during operation. High-speed motors can produce a sirenlike noise if the fan housing is not properly designed. In reassembling a motor, make sure the fan blades do not touch the motor housing. If they do, the noise will be heard right away. Spacing washers are used to prevent end play that causes the fan to hit the motor housing; properly installed, they prevent noise from the fan. Of course, the proper size of washer is important.

Fig. 4–9. To prevent ball-bearing movement, a special spring washer is placed in the bell housing.

Finally, gear trains may or may not be noisy. Worm-gear-type trains are almost noiseless. Helical gearing is also quiet. Spur gearing makes the most noise. Gear-related noise shows up when the load increases. It becomes more intense and evident as the load is increased. Backlash noise occurs in gear trains when the motor is operated at little or no load. Some noise at low speeds and low load is normal for all gear-type trains.

Getting Rid of Mechanical Noises

The first step in getting rid of mechanical noise is to find the source of the noise. To help eliminate noise, (1) enclose the motor to reduce the airborne noise that may come from the inside of the motor housing; (2) use a resilient mounting to help dampen motor vibrations that reach the mounting structure (Fig. 4–10); (3) use flexible couplings to reduce noise; and (4) incorporate rubber, cork, and felt in the mounting design to absorb sound.

Electrical Motor Noise

The hum of a running motor may be music to some people, but it is very annoying to others. The noise itself may be caused by the electrical characteristics of the motor. Such noise is made up of loudness and frequency.

Slip in a motor can cause annoying noise. Slip is the difference between the speed of the rotating field and actual speed of the rotor. Induction motors, including the split-phase, have a lot of slip. They are quieter during operation than reluctance-synchronous motors, which operate with no slip. The amount of slip has a direct bearing on the level of noise generated by the electrical characteristics of a motor.

The *air gap* can also cause noise. The gap between the rotor and the stationary poles is important in noise production. When the air gap

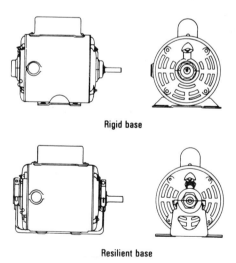

Rigid base

Resilient base

Fig. 4–10. Rigid-base and resilient-base mountings for electric motors.

Table 4–2. Motor Characteristics

Split-Phase AC Motor (nonsynchronous)
Duty: Continuous Power Supply: AC Reversibility: At rest only Speed: Relatively constant Starting Torque: 175 percent and up (of rated torque) Starting Current: High

is too small, the stator teeth can become oversaturated and the motor noise will increase.

The *distribution of the field flux* makes a great deal of difference in the quietness of an electric motor. The quietness of motor operation depends on the strength of the magnetic flux and how it is distributed. For example, the permanent-split capacitor motor has two windings and thus has a more even flux distribution. It is less noisy than the split-phase motor. Troubleshooting and maintenance of the split-phase motor is found in detail in Chapter 9.

CHAPTER 5

Capacitor-Start Motors

The capacitor-start motor is a modified form of the split-phase motor. It has a capacitor in series with the start winding. An external view shows the capacitor sitting on top or on the side of the motor frame (Figs. 5–1 and 5–2). the capacitor produces a greater phase displacement of currents in the start and run windings than is produced in the split-phase motor.

The start winding is made of many more turns of larger wire than in the split-phase motor, and the winding is in series with the capacitor. The start-winding current is displaced 90° from the run-winding current. Since the axes of the two windings are also displaced by an angle of 90°, these conditions produce a greater starting torque than that of the split-phase motor. The start torque of the capacitor may be as much as 350 percent of the full-load torque.

If the start winding is cut out after the motor has increased in speed, the motor is called a capacitor-start motor. If the start winding and capacitor are designed to be left in the circuit continuously, the motor is called a capacitor-run motor. Electrolytic capacitors for capacitor-start motors vary in size from about 80 microfarads (μF) for ⅛-horsepower motors to 400 microfarads for 1-horsepower motors. Capacitor motors of both types are made in sizes ranging from fractional horsepower to about 10 horsepower. They are used to drive grinders,

Fig. 5–1. Capacitor-start motor. Note the location of the start capacitor on the top of the motor. *(Courtesy Westinghouse)*

drill presses, refrigerator compressors, and other loads that require relatively high starting torque. The direction of rotation of the capacitor motor may be reversed by interchanging the start-winding leads. Some table saws are powered by capacitor motors. This provides high pull-out torque that is needed when sawing hard wood or if a knot is hit when ripping through a board.

Starting a Capacitor-Start Motor

The capacitor-start motor is similar to the split-phase motor in starting, with one exception—the insertion of a capacitor in series with

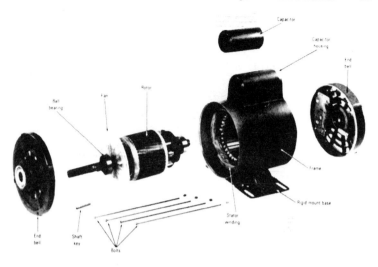

Fig. 5–2. Exploded view of the capacitor-start motor. *(Courtesy Westinghouse)*

the winding. The windings of the motor are displaced 90° electrically. In order to develop a rotating field, it is necessary to develop locked-rotor torque. The currents in the start and run windings must be displaced by 90°. In both the capacitor and split-phase motors, the windings are connected in parallel to the line or power supply. In the split-phase motor, resistance is deliberately built into the auxiliary winding to bring the current more nearly in phase with the line voltage. In a capacitor-start motor, the capacitor causes the start-winding phase current to *lead* the main-phase voltage, obtaining a large angle of displacement between the currents in the two windings. The line current in this motor is only two-thirds of the line current of the split-phase motor. This motor, however, has more than twice the locked-rotor torque of the split-phase motor. The capacitor is more effective as a starting device than the resistance in the split-phase motor. However, it must be remembered that the start current is rather high and must have a fuse or a circuit breaker that will allow the higher current draw from the line for a short period before disconnecting the circuit.

The Electrolytic Capacitor

Most modern-day capacitors used on AC are encased in a black Bakelite container (Fig. 5–3). Many capacitors have two terminals that allow for a slip-on type of solderless wire connector. In other types a screw-type terminal is mounted on the electrolytic. This type of starting arrangement has been in use since 1892. However, it was not until 1930 that the capacitor-start motor was generally accepted as a standard power source for devices which start under load conditions.

The dry electrolytic type capacitor is made by winding two sheets of aluminum foil into a cylindrical shape. The two sheets of aluminum foil are separated by an insulator, which is usually gauze. Two layers of paper are also used in combination with an electrolyte. The electrolyte

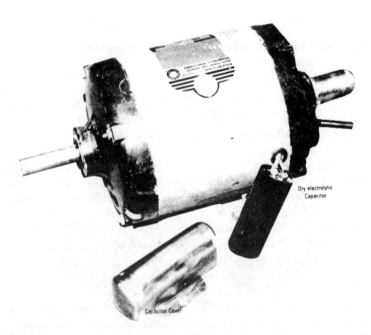

Fig. 5–3. Capacitor-start motor. Note how the Bakelite case of the electrolytic capacitor shows after the metal cover has been removed. A piece of corrugated cardboard is inserted underneath the capacitor to insulate it from the case and reduce noise.

is usually ethylene or some similar chemical composition. On each of the two layers of foil, an anodic film is produced by electromechanical means; that is, the finished capacitor has a DC voltage applied to its terminals. This means it has a positive (+) and a negative (−) marking. The capacitor is sealed in an aluminum case (Fig. 5–4). There are other types of cases, more widely used than aluminum, for the outside of the capacitor.

If AC is applied to an electrolytic that has been formed with a positive (+) and a negative (−) orientation on its terminals, the capacitor will *explode* in about 5 seconds. That is why the AC capacitor is actually two capacitors in one. The two capacitors are connected back to back in series; that is, the positive (+) of one capacitor is connected to the positive (+) of the other capacitor. That produces a *nonpolarized* electrolytic capacitor for use on AC (Fig. 5–5).

Characteristics of the Electrolytic Capacitor

Electrolytic capacitors used on motors are usually of a larger size than those used in electronic circuits for several reasons. One reason is that they have to be able to take the vibration and mechanical abuse of

Fig. 5–4. Electrolytic capacitors used for capacitor-start motors. The round ones are usually used for capacitor-start operations. The other shapes are usually used for capacitor-run operations.

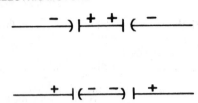

Fig. 5–5. Series connections of capacitors (polarized types) to make them nonpolarized for use on AC motors.

a motor. Another is that they are located in a space that is not too confining and they can be larger to dissipate the heat buildup that occurs in the capacitor during its brief stay in the circuit. Some capacitors are designed to remain in the circuit in the capacitor-run type of electric motors (Fig. 5–6).

The voltage rating is stamped on the outside of the capacitor case. This is very important. It tells you the safe operating voltage for that particular capacitor. It becomes very important when you switch from 120-volt operation of the motor to 240 volts. The capacitor must be able to take the higher voltage, as should the motor windings. However, since most of these motors are made to run on either 120 or 240 volts, the engineers see to it that the capacitor will operate on both. The voltage rating tells what voltage it takes for the capacitor to break down.

About the only problem with normal operation of the capacitor-

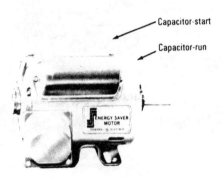

Fig. 5–6. Small motor with two capacitors. Capacitor-start and capacitor-run capacitors are shown mounted on the outside of the motor frame. (Courtesy General Electric)

start motor is with the capacitor. After a few years of use, or even after no use, the capacitor has a tendency to dry out. This causes the capacitor to open. It no longer acts as a capacitor and the motor will not start; it will merely hum. However, if the motor is started manually, it will start in either direction. It will come up to speed and run properly. Replacement of the capacitor should always be with a unit of the same capacitance and voltage rating.

A *farad* is the unit of measurement for capacitance. It is described as the capacitance of a capacitor in which a charge of 1 coulomb produces a change of 1 volt in the potential difference between its plates or terminals. This means a lot of electrons and the one-farad capacitor is huge, especially if the working voltage is 120 or 240. Therefore, it is only necessary for most electronic and electrical applications to use very small amounts of capacitance to get the job done. The term *micro* means one-millionth (0.000001); hence the microfarad is one-millionth of a farad. Eighty microfarads equal eighty-millionths of a farad.

Most electrolytic capacitors will operate in temperatures up to 176°F (80°C). That means they are useful for most motor applications. They are useful for shorter periods of time if they operate at elevated temperatures.

Duty cycle for motor-starting capacitors is rated on the basis of twenty 3-second periods per hour. Sixty 1-second periods per hour would be one equivalent duty cycle. Table 5–1 lists the ratings and test limits for AC electrolytic capacitors. *Caution: When you replace a defective capacitor, it is imperative that the new capacitor be of the same voltage and microfarad rating.*

Characteristics of Capacitor Motors

There are many types of motors which use capacitors in their start circuit, in their start and run circuit, and in combinations of start and run. These motors are single-phase induction motors with main winding arranged for direct connection to the power source. The start winding has a capacitor connected in series with it. There are three types of capacitor motors: the capacitor-start, the permanent-split capacitor, and the two-value capacitor motor. The capacitor-start motor has the capacitor in the circuit only during starting. The permanent-split capacitor motor has the same capacitor and capacitor phase in the circuit

Table 5–1. Ratings and Test Limits for AC Electrolytic Capacitors

Capacity Rating, Microfarads			110-Volt Ratings		125-Volt Ratings		220-Volt Ratings	
Nominal	Limits	Average	Amps. at Rated Voltage, 60 Hz	Approx. Max. Watts	Amps. at Rated Voltage, 60 Z	Approx. Max. Watts	Amps. at Rated Voltage, 60 HZ	Approx. Max. Watts
	25– 30	27.5	1.04– 1.24	10.9	1.18– 1.41	14.1	2.07–2.49	43.8
	32– 36	34	1.33– 1.49	15.1	1.51– 1.70	17	2.65–2.99	52.6
	38– 42	40	1.56– 1.74	15.3	1.79– 1.98	19.8	3.15–3.48	61.2
	43– 48	45.5	1.78– 1.99	17.5	2.03– 2.26	22.6	3.57–3.98	70
50	53– 60	56.5	2.20– 2.49	21.9	2.50– 2.83	28.3	4.40–4.98	87.6
60	64– 72	68	2.65– 2.99	26.3	3.02– 3.39	33.9	5.31–5.97	118.2
65	70– 78	74	2.90– 3.23	28.4	3.30– 3.68	36.8	5.81–6.47	128.1
70	75– 84	79.5	3.11– 3.48	30.6	3.53– 3.96	39.6	6.22–6.97	138
80	86– 96	91	3.57– 3.98	35	4.05– 4.52	45.2	7.13–7.96	157.6
90	97–107	102	4.02– 4.44	39.1	4.57– 5.04	50.4	8.05–8.87	175.6
100	108–120	114	4.48– 4.98	43.8	5.09– 5.65	56.5	8.96–9.95	197
115	124–138	131	5.14– 5.72	50.3	5.84– 6.50	65		
135	145–162	154	6.01– 6.72	62.8	6.83– 7.63	85.8		
150	161–180	170	6.68– 7.46	69.8	7.59– 8.48	95.4		
175	189–210	200	7.84– 8.71	81.4	8.91– 9.90	111.4		
180	194–216	205	8.05– 8.96	83.8	9.14–10.18	114.5		
200	216–240	228	8.9⌂– 9.95	93	10.18–11.31	127.2		
215	233–260	247	9.66–10.78	106.7	10.98–12.25	145.5		
225	243–270	257	10.08–11.20	110.9	11.45–12.72	151		
250	270–300	285	11.20–12.44	123.2	12.72–14.14	167.9		
300	324–360	342	13.44–14.93	147.8	15.27–16.96	201.4		
315	340–380	360	14.10–15.76	156				
350	378–420	399	15.68–17.42	172.5				
400	430–480	455	17.83–19.91	197.1				

for both starting and running. The two-value capacitor motor has capacitors of different capacitance values for starting and for running.

The *permanent-split capacitor motor* is suitable for operation only on AC. Its standard windings are designed for continuous duty. This type of motor may be reversed at standstill or, under favorable conditions, when running. High-inertia loads retard the reversing action. Friction loads present no problems with reversing.

This permanent-split capacitor motor has fairly constant speed, as does the split-phase motor. It does not have low starting torque, though, and pulls very little current on starting. It is the quietest-operating and smoothest-performing of all the classes of single-phase fractional-horsepower motors.

The *capacitor-start motor* is suitable for operation only on AC. Its windings are designed for continuous duty. It may be reversed at standstill or when rotating at a speed low enough to ensure that the start winding is in the circuit. The motor can be reversed at full speed by the use of a special type of centrifugal switch or by a relay.

The *two-capacitor motor* (Fig. 5–6) is suitable for operation only on AC. The windings are also designed for continuous duty. Its starting and reversing characteristics are similar to those of the capacitor-start motor. It has less breakdown torque than the polyphase motors, but has similar run characteristics.

Reversing the Motor

An induction motor will not always reverse while running. It may continue to run in the same direction, but at a reduced efficiency. An inertia-type load is difficult to reverse. Most motors that are classified as reversible while running will reverse with a noninertia-type load. They may not reverse if they are under no-load conditions, have a light load, or an inertia load. A permanent-split capacitor motor that has insufficient torque to reverse a given load may just continue to run in the same direction.

One of the problems related to the reversing of a motor while it is still running is the damage done to the transmission system connected to the load. In some cases it is possible to damage a load. One of the ways to avoid this is to make sure the right motor is connected to a load.

Dual-Frequency Operation

Capacitor-start motors are least suitable for the dual-frequency operation. However, it may be possible in some cases.

The permanent-split capacitor motor is best suited for 50/60-Hz operation. This is due primarily to the two different effects on the two windings with a change in frequency. When the frequency is changed from 60 to 50 Hz, the current in the main winding will decrease. This

Table 5–2. Motor Characteristics

Capacitor Motor
Duty: Continuous Power Supply: AC Reversibility: At rest or during rotation Speed: Relatively constant Starting Torque: 75–150 percent of rated torque Starting Current: Normal

means that the total current may remain the same. It is possible to wind a permanent-split capacitor motor so that it will not draw more power from the line at one frequency than the other. However, other operating characteristics may change. It is best to know under what conditions the motor will operate and then select the proper motor for doing the job.

Mechanical and Electrical Noises

The capacitor motors have about the same types of mechanical noises as the split-phase motor (see Chapter 4). They are essentially built the same mechanically, with the addition of a start capacitor mounted to the case in most applications. In the case of a refrigerator motor, the capacitor is mounted on the inside wall of the refrigerator cabinet or under the main unit. In the case of a washing machine, it is usually mounted on a removable panel in the rear of the machine. This means that it does little to add to the mechanical noise level.

Electrical noises are almost identical to those generated by the split-phase motor (see Chapter 4). However, most of the capacitor motors are smoother-running than the split-phase motors, and they have a lower noise level than other types of single-phase motors.

Most of the noise can be eliminated by using a resilient type of mounting for the motor and by checking closely the location of the motor in reference to the device it is powering.

CHAPTER 6

Shaded-Pole Motors

The shaded-pole motor is a single-phase induction motor with an auxiliary short-circuited winding or windings. The shorted winding is displaced in magnetic position from the main winding (Fig. 6–1). Shaded-pole motors are suitable for operation on AC only. The standard windings are designed for continuous duty when used to power fans that will draw air over the windings. The shaded-pole motor is not normally reversible. It takes a stator with two windings wound differently to make it reversible (Fig. 6–2). The speed of a shaded-pole motor is fairly constant, but it will vary once a load is placed on it. It has very low starting torque—less than 100 percent in most cases. Starting current, however, is also very low—less than twice full-load current (Table 6–1).

Starting the Shaded-Pole Motor

Just as with any other motor, when the main field coils in a shaded-pole motor are energized, a magnetic field is set up between the pole pieces and the rotor. A portion of the magnetic field is cut by a *shading* coil (Fig. 6–3). This coil, made of a large-diameter piece of wire shorted together in the form of a ring, shades the pole piece of the

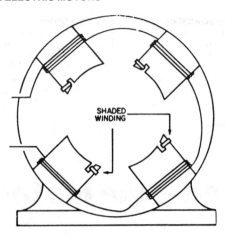

Fig. 6–1. Four-pole motor stator showing the shaded windings.

main windings. This shorted winding makes the magnetic field between the pole piece and rotor slightly out of phase with the main portion of the pole piece. This has the effect of producing a two-phase voltage situation at every pole piece. There is very little torque generated by this phase difference, but it is enough to get the motor started if it is not overloaded. The rotor is a squirrel-cage configuration, which has a high resistance. The net effect of the shading of the pole piece is to produce a displacement that causes a shifting flux in the air gap, always shifting toward the shading coil. The direction of rotation of a shaded-pole motor is always from the unshaded to the shaded portion of the pole.

Fig. 6–2. Dual stator for reversibility. *(Courtesy Brevel)*

Table 6–1. Characteristics of the Shaded-Pole Motor

Characteristic	Rating	Comment
Efficiency	Low	20–40 percent
Power factor	Low	50–60 percent
Starting torque	Low	Plus 3rd harmonic dip
Noise and vibration	High	120 Hz plus winding harmonics
Cost	Low	

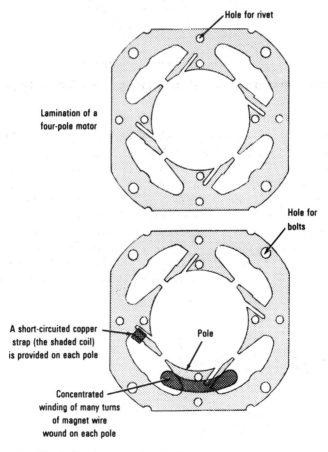

Fig. 6–3. Shaded-pole motor laminations.

Types of Shaded-Pole Motors

Shaded-pole motors are of three basic types and are built in a wide variety of ratings, with any number of performance characteristics. They are made in subfractional and fractional horsepower only. Horsepower ratings vary from 25 hundred-thousandths (0.0025) to one-tenth (0.1) horsepower. Some serious design characteristics prevent them from being made in larger horsepower sizes. In the upper horsepower range, most of the motors are of the four-pole construction method (Fig. 6–4), while the lower-horsepower types use only two poles.

Speeds for the shaded-pole motor are stated with no-load conditions, and range from 2700 to 3500 rpm. Full-load speeds are from 1600 to 3000 rpm. The most common 60-Hz types with four poles have a speed of 1750 to 1770 rpm. Full-load speeds are then 1450 to 1620 rpm. A good many of these motors are designed with a reduction gearbox attached. They can be reduced to as much as 450:1, which means that a wide variety of speeds is available for use in many places. General Electric makes four-pole shaded-pole motors in 1000, 1100, and 1500 rpm.

Fig. 6–4. Four-pole shaded-pole motor. *(Courtesy General Electric)*

Construction of Shaded-Pole Motors

The three types of shaded-pole motors are salient-pole, skeleton, and distributed-winding. *Salient-pole* construction has many main-winding coils. The number of poles is the same as the number of coils. In most cases there are four coils. Some of the older refrigerator fans are salient-pole shaded-pole motors. They used a felt soaked in oil for lubricating the rotary part.

The *skeleton* type is used for horsepowers from 0.00025 to 0.03 (Figs. 6–5 and 6–6). This type of motor uses triple shading. Three shading coils of different throw for each pole are used. In this type the bearings are self-aligning oilite. Wick-type oilers are used to spread the lubrication so that it covers the whole rotating shaft. A squirrel-cage rotor is used in the skeleton design. A spring causes the rotor to be slightly offset from the pole, and it is easily compressed as the coil is energized. The sucking effect of the coil causes the rotor to be pulled

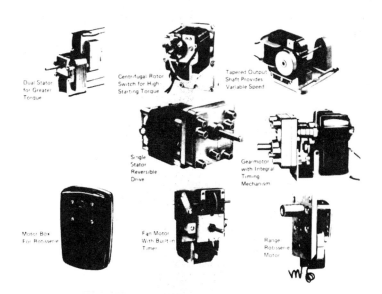

Fig. 6–5. Different types of special-design shaded-pole motors.
(Courtesy Brevel)

Fig. 6–6. Typical bearing brackets for skeleton-type shaded-pole motors. *(Courtesy Brevel)*

back inside the hole in the laminated core. This type of motor is used on can openers, small knife sharpeners, clocks, and timers. It has an advantage over the synchronous type originally used in clocks, as it is self-starting and will start again if the power fails. The old clock-type synchronous motors had to be started by hand.

The third type of construction of shaded-pole motors is the *distributed-winding*. The stator laminations are similar to those used for single-phase or polyphase induction motors. The main winding is also similar to the single-phase motor. The short-circuited auxiliary winding is similar to the start winding except that it is short-circuited upon itself. It is displaced from the main winding by less than the 90° usually found in induction motors.

Performance of Shaded-Pole Motors

If this type of motor is properly lubricated—in most cases the skeleton type is sealed—it will last (continuously operated) for over 25 years. Most clock motors and fan motors are limited only by the physical abuse they receive. If they are kept plugged in (in the case of a clock) and operating continuously as timers or similar devices, without overloading, there is no reason why they cannot last indefinitely. There is only one coil in the skeleton-type motor. If the voltage is stable and the temperature is normal, there is no reason why it will not continue to operate without maintenance of any kind. Various types of physical and electrical abuse can cause them to fail, however.

If the timer motor has been used to power the timing mechanism of an oven, then it will be only a matter of eight to ten years of sporadic use that will cause it to lose its lubrication. The heat from the oven can cause it to lose its oil and the seals to break down.

Applications of Shaded-Pole Motors

The efficiency of a shaded-pole motor suffers due to the presence of winding *harmonics.* A dip is noticed in the speed-torque curve. It is caused by the third harmonic at approximately one-third synchronous speed. The shaded-pole is the least efficient and the noisiest of the single-phase motors. It is used mostly in air-moving applications where its low starting torque and third harmonic dip can be tolerated. Extra main windings can be added to provide additional speeds in a manner similar to that used on permanent split-capacitor motors. This is the case in the 1000-, 1100-, and 1500-rpm three-speed designs. This type of motor is used in portable fans, space heaters, forced-draft types of heaters, ventilators, room air conditioners, recorders, humidifiers, and range hoods.

Fig. 6–7 illustrates a four-pole shaded-pole motor with a double shaft. In some cases you may want to buy this type and cut off one end to make it fit the needed application. The mounting screws are long so they too can be cut to fit any type of replacement requirement.

Diagram connections for the General Electric shaded-pole motors are shown in Fig. 6–8. Note the two-speed and three-speed color codes. Most companies have a standard color code for their wiring or hookups.

Fig. 6–9 shows another manufacturer's type of four-pole shaded-

Fig. 6–7. Double-shaft four-pole shaded-pole motor for replacement of damaged fan motors. *(Courtesy General Electric)*

CONNECTION DIAGRAMS

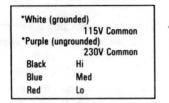

*White (grounded)
          115V Common
*Purple (ungrounded)
          230V Common

Black	Hi
Blue	Med
Red	Lo

* Caution: Using any other col-
  or than white or purple as
  common will burn out motor.

KSM59 Shaded Pole (General Electric)

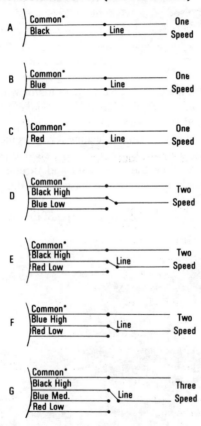

Fig. 6–8. Connection diagrams for General Electric shaded-pole replacement motors. *(Courtesy General Electric)*

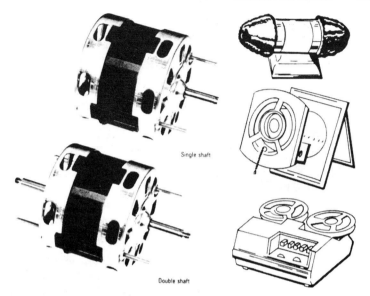

Single shaft

Double shaft

Fig. 6–9. Shaded-pole motors—four-pole, 1550 rpm, 115 volt, 60 Hz.
(Courtesy Voorias)

pole motor. This is a standard four-pole 1550-rpm, 115-volt, 60-Hz design. It is used in the usual air-moving jobs and in the shoe polisher and the tape recorder. It is designed for face mount with two #8-32 case screws. It has sleeve bearings and an oil hole to keep it properly lubricated. This type of motor needs little maintenance except for a few drops of oil at regular intervals.

There are a number of types and shapes attached to the shaded-pole motor. In Figs. 6–10, 6–11, and 6–12, you can see some of the wide variety of shapes and designs. Note which are the two-pole, which are the four-pole, and which are the six-pole; it is possible to identify them without taking them apart and checking the coil. All of these motors use sleeve bearings. The only maintenance needed is oiling and keeping the surface area around the coils free of dust and dirt that may cause the rotor to jam.

The skeleton shaded-pole motor has a very limited number of parts. It has only one coil, unless it is used in a special application as shown in Fig. 6–5. This motor may be used to drive inexpensive turn-

Fig. 6–10. Two-pole shaded-pole motors. *(Courtesy Fasco)*

Fig. 6–11. Four-pole shaded-pole motors. *(Courtesy Fasco)*

tables or for clocks. The horsepower rating will vary with the size of the laminations, coil, and rotor.

Fig. 6–13 shows a skeleton-type shaded-pole motor. This one was used to power an inexpensive record player—that is why the shaft has varying diameters.

In Fig. 6–14 you can see the rotor inside the motor lamination. The top bracket and bearing have been removed so you can see the rotor in place.

Fig. 6–15 shows the entire motor disassembled. The coil is still on the frame. There are two spacing washers and a spring in addition to the frame. The end caps or bearings have an oil-retaining wick to keep the shaft lubricated while it rotates. About the only maintenance necessary is oiling once in many years of operation. If it has operated in high temperatures or has a humidity problem surrounding it, as when it is used for a timer clock on an electric oven, you may have to lubricate it more often or check for corrosion in some cases.

In some cases the shaded-pole motor has been designed to do specific jobs. One of those special jobs is the inexpensive grinder, buffer, or a similar type of tool. K-Mart, Black & Decker, and some other 6-inch grinders are powered with a shaded-pole motor. Note the shading coil on the motor in Fig. 6–16. It has only two windings; they are connected in series and to a switch and continue out to a plug. This type of motor is sufficient for low-usage or home-workshop applications. It is

Fig. 6–12. Six-pole shaded-pole motors. *(Courtesy Fasco)*

Fig. 6–13.   Skeleton-type shaded-pole motor.

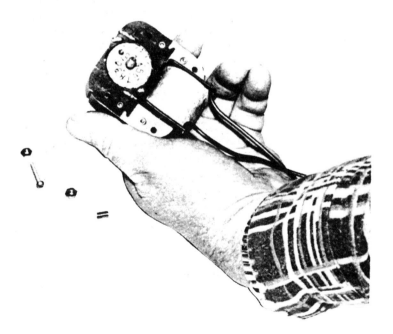

Fig. 6–14.   Skeleton-type with one bearing and bracket removed.

Fig. 6–15. All the parts of a skeleton-type shaded-pole motor.

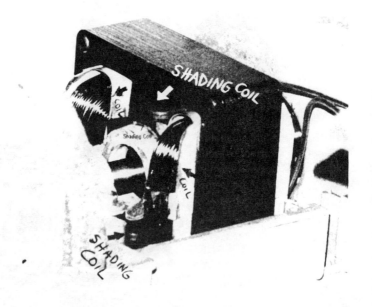

Fig. 6–16. Close-up view of a shaded-pole motor used in a grinder-buffer. Note the arrows show the shading coil.

not sufficient for use in production. Production machines have a split-phase motor to power the grinders, wire brushes, and buffers.

As the price of shaded-pole motors continues to decrease in comparison with other types, you will probably find more of this size of shaded-pole motor in machines that once used split-phase or similar motors. This one is inexpensive and has no starting mechanism.

CHAPTER 7

Three-Phase Motors

Most industrial motors are three-phase. The main reason for this is that the maintenance of a three-phase motor is practically nil. Industrial motors do not have the starting devices that single-phase motors have. The three phases of AC that supply power for the motor produce the phase shift needed to get the motor started and to keep it running once it is started. Fig. 7–1 shows the three phases of AC as they appear in a graphic form. Notice how each phase is displayed 120° from the other. All commercial power generated in the United States is generated as three-phase. It is converted to single-phase because the three separate phases can be divided and sent into three different subdivisions or locations. It is cheaper to distribute the single-phase AC than three-phase AC. Three-phase power requires at least three, and sometimes four, wires for proper distribution.

Uses of Three-Phase Motors

These three-phase motors (3ϕ) are ideal for machine-tool and general uses where dust and dirt are prevalent. Polyphase motors have operating characteristics which enable them to operate any device that may be powered by equivalently rated single-phase motors. The three-

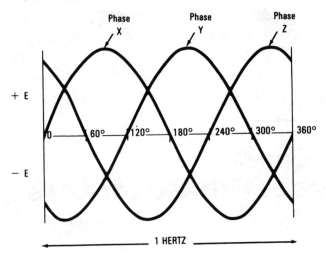

Fig. 7–1. Three-phase alternating current.

phase motors you see in Fig. 7–2 are available in sizes of ¼, ⅓, ½, and ¾ horsepower. They may be used for pumps, compressors, fans, blowers, conveyers, farm machinery, saws, and machine tools.

The motors shown in Fig. 7–2 have ball bearings with permanent lubrication. They are cooled by a fan and are fully enclosed, with a service factor of 1.0.

Look at Table 7–1 and see how the size of the motor increases with horsepower. The ¼-horsepower motor weighs 19 pounds; the ¾-horsepower motor weighs 30 pounds. Speeds are 850, 1140, 1725, 1425, and 3450 rpm.

Also available are three-phase motors with a nonventilated construction (Fig. 7–3). They are totally enclosed to eliminate cleaning and clogging problems, and are well suited for operation in the lint-laden atmospheres of the textile industry and other industrial areas having similar environmental conditions.

This type of motor is built to handle heavy thrust loads. The grease used in the ball bearings is resistant to oxidation and moisture. These motors should last for ten years under normal operating conditions. They are rated for continuous duty in temperatures up to 108°F (40°C). These motors are made by General Electric in the ¼-, ⅓-, ½-, and

NEMA 48-Frame
Rolled Steel Construction

NEMA 56H-, 143T, 145T-Frame,
Rolled Steel Construction

NEMA 143T- to 184T-Frame,
Cast-aluminum Construction

NEMA 213T- 326T-Frame,
Cast-aluminum Construction

Fig. 7–2. Fully enclosed three-phase motors.

$\frac{3}{4}$-horsepower size. They have a speed of 1725 rpm. They weigh from 15 lb. for the $\frac{1}{4}$-horsepower motor to 33 pounds for the $\frac{3}{4}$-horsepower motor.

Explosion-proof three-phase motors are used in hazardous locations where a spark could start a fire or cause an explosion. The explosion-proof motor is ideal for use in locations such as dry-cleaning and dyeing plants, paint and varnish factories, alcohol and acetone plants, gasoline refineries, hospitals and laboratories, flour and feed mills, grain elevators, starch, sugar, coke or coal plants, and other locations requiring motors which are UL listed for hazardous locations. Fig. 7–4 shows two explosion-proof motors.

How Three-Phase Motors Work

The stator has windings around it that are placed 120° apart. The rotor is a form-wound type or a cage type. The squirrel-cage rotor is standard for motors smaller than 1 horsepower, which we are concerned with here. Fig. 7–1 shows how the three phases are produced

Table 7–1.   Three-Phase Motor Weight and Horsepower

HP	Speed RPM	Volts	NEMA Frame	Bearings	Therm. Prot.	F.L. Amps @230V	Est. Shpg. Wt. (Lbs.)
¼	1725	230/460	48	Ball	None	1.5	19
	1140	230/460	56	Ball	None	1.3	22
	850	230/460	56	Ball	None	2.0	27
⅓	3450	230/460	48	Ball	None	1.7	19
	1725	230/460	56	Ball	None	1.5	22
	1140	230/460	56	Ball	None	1.7	25
½	3450	230/460	48	Ball	None	2.0	21
		230/460	56	Ball	None	2.0	21
	1725	230/460	56	Ball	None	2.2	22
	1425	220/380	56	Ball	None	2.3	20
	1140	230/460	56	Ball	None	2.6	27
¾	3450	230/460	56	Ball	None	2.6	24
	1725	200	56	Ball	None	3.6	24
		230/460	56	Ball	None	3.2	24
	1425	220/380	56	Ball	None	2.6	27
	1140	230/460	56	Ball	None	3.1	30
		230/460	143T	Ball	None	3.1	27

(Courtesy General Electric)

by the generator. The rotor will rotate with the rotating field produced by the stator. The stator is nothing more than the primary of a three-phase transformer. The magnetic field produced by the stator revolves and cuts across the rotor conductor. This induces voltage and causes the rotor current to flow. Hence, motor torque is developed by the in-

Fig. 7–3.   Totally enclosed, non-ventilated three-phase motor.

NEMA 56-Frame

Fig. 7–4. Explosion-proof, fan-cooled three-phase motor.

NEMA 145T- through 286T- Frame

teraction of the rotor current and the magnetic revolving field. Fig. 7–5A and B show how the field rotates. The large motor stator and rotor (Fig. 7–5C) are shown here to illustrate the details a little more clearly for the large industrial type of three-phase motor.

The purpose of the iron rotor is to reduce the air-gap reluctance and to concentrate the magnetic flux through the rotor conductors. Induced current flows in one direction in half of the rotor conductors, and in the opposite direction in the remainder. The shorting rings on the ends of the rotor complete the path for rotor current. In Fig. 7–5D, a two-pole field is assumed to be rotating in a counterclockwise direction at synchronous speed. At the instant pictured, the south pole cuts across the upper rotor conductors from right to left, and the lines of force extend upward. Applying the left-hand rule for generator action to determine the direction of the voltage induced in the rotor conductors, the thumb is pointed in the direction of motion of the conductors with respect to the field. Since the field sweeps across the conductors from right to left, their relative motion with respect to the field is to the right. Hence, the thumb points to the right. The index finger points upward and the second finger points into the page, indicating that the rotationally induced voltage in the upper rotor conductors is away from the observer.

Motor action is analyzed by applying the right-hand rule for motors to the rotor conductors in Fig. 7–5D, to determine the direction of

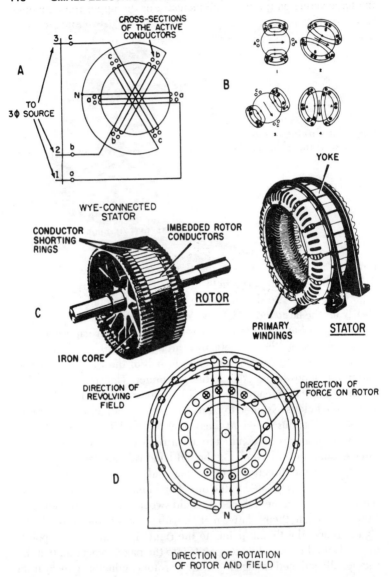

CROSS-SECTIONS OF THE ACTIVE CONDUCTORS

A

TO 3Φ SOURCE

WYE-CONNECTED STATOR

B

CONDUCTOR SHORTING RINGS

IMBEDDED ROTOR CONDUCTORS

C

ROTOR

IRON CORE

YOKE

PRIMARY WINDINGS

STATOR

DIRECTION OF REVOLVING FIELD

DIRECTION OF FORCE ON ROTOR

D

DIRECTION OF ROTATION OF ROTOR AND FIELD

Fig. 7–5. Reversing switch for a three-phase reversing motor.
(Courtesy Square D)

the force acting on the rotor conductors. For the upper rotor conductors, the index finger points upward, the second finger points into the page, and the thumb points to the left, indicating that the force on the rotor tends to turn the rotor counterclockwise. This direction is the same as that of the rotating field. For the lower rotor conductors, the index finger points upward, the second finger points toward the observer, and the thumb points toward the right, indicating that the force tends to turn the rotor counterclockwise—the same direction as that of the field. Fig. 7–6 illustrates the right-hand rule. Take a close look at Fig. 7–7, where the rotor is shown in a cutaway view of the rest of the motor.

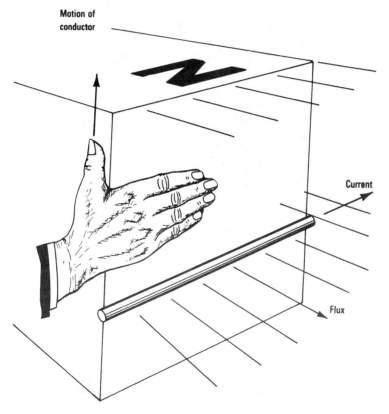

Fig. 7–6. The right-hand rule for motors.

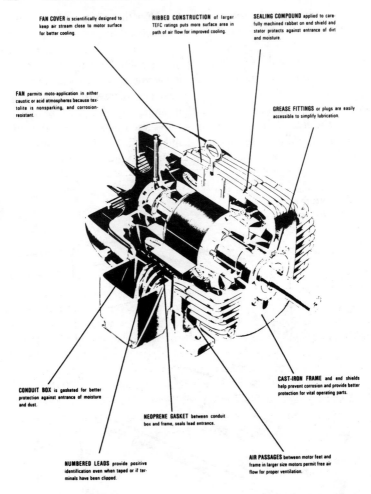

FAN COVER is scientifically designed to keep air stream close to motor surface for better cooling.

RIBBED CONSTRUCTION of larger TEFC ratings puts more surface area in path of air flow for improved cooling.

SEALING COMPOUND applied to carefully machined rabbet on end shield and stator protects against entrance of dirt and moisture.

FAN permits moto-application in either caustic or acid atmospheres because textolite is nonsparking, and corrosion-resistant.

GREASE FITTINGS or plugs are easily accessible to simplify lubrication.

CONDUIT BOX is gasketed for better protection against entrance of moisture and dust.

CAST-IRON FRAME and end shields help prevent corrosion and provide better protection for vital operating parts.

NEOPRENE GASKET between conduit box and frame, seals lead entrance.

NUMBERED LEADS provide positive identification even when taped or if terminals have been clipped.

AIR PASSAGES between motor feet and frame in larger size motors permit free air flow for proper ventilation.

Fig. 7–7. Cutaway view for the totally enclosed three-phase motor. *(Courtesy General Electric)*

The stator of a polyphase (3ϕ) induction motor consists of a laminated steel ring with slots on the inside circumference. The motor winding is similar to the AC generator stator winding and is generally of the two-layer distributed, preformed type. Stator phase windings are symmetrically placed on the stator and may be either wye or delta connected.

Torque

The revolving field produced by the stator windings cuts through the rotor conductors and induces a voltage in the conductors. Rotor currents flow because the rotor end rings provide continuous metallic circuits. The resulting torque tends to turn the rotor in the direction of the rotating field. This torque is proportional to the product of the rotor current, the field strength, and the rotor power factor.

By using the transformer comparison, it is possible to see that the primary is the stator and the secondary is the rotor. At start, the frequency of the rotor current is that of the primary stator winding. The reactance of the rotor is relatively large compared with its resistance, and the power factor is low and lagging by almost 90°. The rotor current therefore lags the rotor voltage by approximately 90°. Because almost half of the conductors under the south pole carry current inward, the net torque on the rotor as a result of the interaction between rotor and the rotating field is small.

As the rotor comes up to speed in the same direction as the revolving field, the rate at which the revolving field cuts the rotor conductors is reduced and the rotor voltage and frequency of rotor currents are correspondingly reduced. Hence, at almost synchronous speed the voltage induced in the rotor is very small. The rotor reactance, X_L, also approaches zero, as may be seen by the relationship

$$X_L = 2\pi f_0\, LS$$

where

f_0 is the frequency of the stator current,
L is the rotor inductance,
S is the ratio of the difference in speed between the stator field and the rotor to the synchronous speed—slip.

Slip is expressed mathematically as

$$S = \frac{N_s - N_r}{N_s}$$

where

N_s is the number of revolutions per minute of the stator field,
N_r is the number of revolutions per minute of the rotor;
the frequency of the induced rotor current is f_oS.

Normal operation is between the time when the rotor is not turning at all and when it is turning almost at synchronous speed. The motor speed under normal load conditions is rarely more than 10 percent below synchronous speed. At the extreme of 100 percent slip, the rotor reactance is so high that the torque is low because of low power factor. At the other extreme of zero rotor slip, the torque is low because of low rotor current.

Motor Ventilation

Motors are commonly ventilated by placing a fan on the rotor. The fan may be at both ends of the motor, or just one fan may be used. Fig. 7–8 shows how the air is taken in at one end of the motor and expelled at the other. Another type of construction is shown in Fig. 7–9. Here the air is taken in at each end of the motor and is circulated over the end coils and exhausts through the openings in the body. Air does not flow from one end to the other in this type of ventilation.

Some losses are inherent in motor design. Fig. 7–10 shows some of the losses due to the construction of the motor, the ventilation method used, and such natural things as eddy current losses, hysteresis losses,

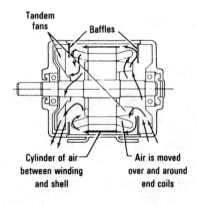

Fig. 7–8. Flow-through ventilation of a motor.

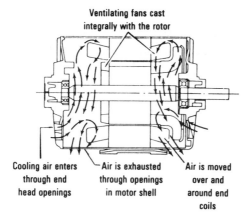

Ventilating fans cast
integrally with the rotor

Cooling air enters
through end
head openings

Air is exhausted
through openings
in motor shell

Air is moved
over and
around end
coils

Fig. 7–9. Ventilation method by way of cooling air flowing through the intake and out the motor shell.

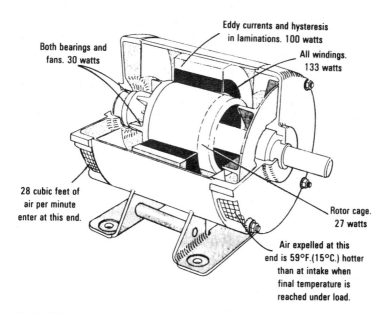

Eddy currents and hysteresis
in laminations. 100 watts

Both bearings and
fans. 30 watts

All windings.
133 watts

28 cubic feet of
air per minute
enter at this end.

Rotor cage.
27 watts

Air expelled at this
end is 59°F.(15°C.) hotter
than at intake when
final temperature is
reached under load.

Fig. 7–10. Ventilation and losses of a three-phase motor.

and copper losses. Eddy currents are kept to a minimum by laminating the stator pole pieces. Hysteresis losses are kept to a minimum by using silicon steel for the core or pole pieces. Copper losses are kept down by using the proper size of wire for the current needed.

Motor Losses

Eddy currents are caused by each piece of metal that will conduct electricity having a small current induced in it from the moving magnetic field. If the piece of metal in the magnetic field is divided into small wafers with insulation between them (in this case a lacquer is usually placed on the lamination sides), then the small pieces of metal offer a larger resistance to eddy currents than a larger piece with low resistance. Eddy currents can cause excess heat in a motor.

Hysteresis losses are caused by the polarity of the material in the pole pieces being magnetized first in one direction and then in the other (in the case of 60 Hz, this means a change of polarity 120 times per second). A material must be used that can easily change its magnetic polarity without undue opposition. After about 30,000 polarity changes per second nothing can keep up, so an air core is used in the coils instead of iron. Of course, this air core is not used with motors in radio frequencies for broadcasting purposes.

Silicon steel for years has been the proper material to use in relays, transformers, and motors in order to keep down the losses due to hysteresis.

Copper losses are simply caused by the size of the wire. The wire has a definite resistance. If you choose the proper size for the current being handled, it should not cause any undue losses. There is no way to effectively eliminate this since the copper wire is needed to cause the current to flow and produce a magnetic field.

Reversibility

Three-phase motors can be reversed while running. It is very hard on the bearings and the driven machine, but it can be done by *reversing any two of the three connections*. This is usually done by a switch specifically designed for the purpose.

Open Phase

If a three-phase motor develops an open "leg" on one phase—that means two instead of three wires are coming into the motor terminals with power—it will slow down and hum noticeably. It will, however, continue to run in the same direction. If you try to start it with only two legs (or phases), it will not start but will rotate if started by hand (in fact, it will start in either direction). Once the other phase is connected, it will quickly come up to speed.

The loss of one leg is usually due to a blown fuse in that leg. That is, of course, if there are three individual fuses in the three-phase circuit.

Fig. 7–11 shows how the three-phase motor is reversed with a drum switch. The handle on the switch can be moved from FORWARD to REVERSE at any time. The motor will come to a complete stop and then start in the opposite direction when the top two lines are switched. The bottom leg of the power line stays the same in both forward and reverse direction.

Applications of Three-Phase Motors

Three-phase motors are used for machine tools, industrial pumps and fans, air compressors, and air-conditioning equipment. They are recommended wherever polyphase power supply is available. They

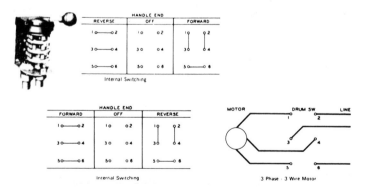

Fig. 7–11. Three-phase drum switch made for reversing the motor rotational direction. *(Courtesy Square D)*

Fig. 7–12. Three-phase motor with a rigid mount. *(Courtesy Leeson)*

provide high starting and breakdown torque with smooth pull-up torque. They are efficient to operate and are designed for 208-230/460-volt operation, with horsepower ratings from ¼ to the hundreds. They can be obtained for 50-Hz as well as 60-Hz operation. Usually, the leads are 6 inches and can be connected in a number of methods if six of them are brought out to a terminal or left in a conduit box. Fig. 7–12 is a totally enclosed, fan-cooled (TEFC) type of three-phase motor with a rigid mount. Fig. 7–13 is a 143T- and 145T-frame three-phase motor. It is used on hard-to-start equipment. Note the difference in the construction of the end bells. They can be bought with a resilient mount as well as with the rigid mount shown.

Fig. 7–13. Three-phase motor, totally enclosed, fan-cooled. *(Courtesy Doerr)*

CHAPTER 8

Special Types of Motors

The adjustable-speed motor, the dust-tight motor, the explosion-proof motor, the high-slip motor, the instant-reversing motor, the farm-duty motor, and the air-conditioner fan motor are all examples of special motors. In some cases three-speed and four-speed motors are also special types. All special motors should not be operated at values other than those on the nameplate. Improper usage accounts for much of the repair or operational problems in electric motors. Look over the following before proceeding.

Understanding the Nameplate

Every motor has a nameplate (Fig. 8–1). This nameplate identifies the motor according to *type*. The motor type is usually coded by the manufacturer. In many cases you almost need the manufacturer's catalog to identify the specifics of the motor. However, by observation and what you have learned so far in this book you should be able to determine some of the characteristics of the motor. The *NO* on the nameplate usually refers to the manufacturer's order or catalog number. Volts will usually be 120, 220, 208. Industrial applications will show

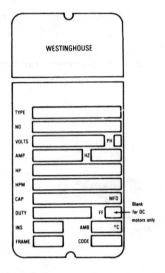

Figure 8–1. Nameplates on motors look about the same as this one. Some slight variations do appear. *(Courtesy Westinghouse)*

440 and 600 volts. The *PH* stands for phase. It will be either single-phase (1) or three-phase (3).

AMP refers to the amperes the motor draws under normal operating conditions. The *HZ* is the frequency of the power needed to properly operate the motor; it will be either 50 or 60 (some very old motors will have 25). Aircraft motors use 400 Hz for power generated on board. Some older nameplates will have *cycle* instead of *Hz* or *Hertz*.

The rpm (*RPM*) may be 1735 or 3450, depending upon the motor and its capabilities. In some cases the speed may be something other than these two. They may also be designated as *1140/1060/990/920/, 10/60/925/840/775*, or *10/75/1000/960/935*. This means that the motor is wound for five different speeds. The leads must be identified in order to obtain the proper speed by switching. Most of these multi-speed motors are permanent split-capacitor types.

The marking *CAP* means the capacitor size. It will be stamped in microfarads for the motor. If this is a split-phase motor, it will not, of course, have a capacitor. *MFD* stands for microfarads.

The word *DUTY* means it is designed for continuous (*CONT*) or intermittent operation (*INT*).

The letters *FF* on the nameplate stand for "form factor." This is a measure of departure from pure DC. It is defined as the root-means-square (RMS) value of the current divided by the average value of the current. For half-wave rectified current, the form factor is 1.57. For full-wave rectified current, the form factor is 1.11. Pure DC has a form factor of 1.0, or unity. FF applies to DC motors only.

Form factor is an important consideration with motors designed to operate on direct current (DC). When operated from rectified power versus pure DC, the increase in motor heating for a constant output is approximately proportional to the square of the form factor. For example, a motor operating from half-wave rectified DC will have approximately 2½ times the heat rise of the same motor operating on unit form factor DC.

In order to accommodate the increased heating effect of high form factor current, continuous-duty applications generally require a larger (and more costly) motor to drive a given load. Stated another way, a designer may save by using a low-cost-to-high-form-factor control, but will have to give up much of the savings by using a larger motor to keep operating temperatures of the motor within design limits.

The letters *INS* stand for the insulation used for the motor wiring. There are three classes of insulation that may appear in this box: Class A is good for 221°F (105°C); Class B for 266°F (130°C); Class F for 311°F (155°C). This is the maximum temperature at which the motor should be operated. This is normally based on a maximum ambient of 40°C (104°F), but one can expect these windings to get hot since higher insulation ratings are used. If the motor is overloaded, it can be expected that the windings' temperature rise will increase and the windings will overheat. Lack of air intake, obstructions to the ventilation flow, and excessive deviations from the nameplate parameters will result in a higher temperature rise. Operating at high temperatures will shorten the life of the motor since it affects the insulation and the lubrication of the bearings.

The letters *AMB* are used to indicate the temperature in which the motor is expected to operate. AMB means ambient.

Other Types of Nameplates

General Electric and Westinghouse each has a different style of nameplate. Other manufacturers of motors also have nameplates that are different in shape and in other ways as well. However, the information is basically the same. In Fig. 8–2 you will find an example of a nameplate for a GE capacitor-start AC motor. This can be interpreted as follows: "HP ¼" means that horsepower is ¼. "FR 48" refers to the frame number, which is 48. "MOD 5KC35KG 145" means the model number can be found in the GE catalog under this number. Pertinent information will be given in the catalog for this particular model. "RPM 1725" gives the revolutions per minute of the motor, 1725. "PH 1" means that the phase is single. "219500" is the motor serial number.

Figure 8–2. General Electric nameplate.

"SF 1.35" is the service factor. "Temp. Rise 40°C" says that the motor can be 40°C above the temperature of the room in which it is operating and still be operating normally. "V 115" is the line voltage for this type of motor. "CODE" is not filled in but it usually means the type of motor designated by the manufacturer. "A 5.2" current is needed to produce the ¼ horsepower. "CY" is frequency of 60 Hertz (new term is "HZ"). "Time Rating CONT" describes the motor as rated for continuous duty. "SER. No. ZRD-2" is the service number.

Westinghouse has a slightly different nameplate on its latest models (Fig. 8–3). The serial number ("S#") is 313P156. Catalog number is AD77 (SER). "HP 2" means that it has two horsepower. The type is FJ. FJ means it is a single-phase motor. RPM or the speed is 1725 rpm. Service factor ("SF") is 1.15. V or voltage required for this motor can be either 115 or 230. The "PH1" means that the phase is single. "A" is used to indicate that at 115 volts the current drawn would be 26 amperes and at 230 volts the current would be 13 amperes. As the voltage is doubled, the current is halved. This is simple to understand when you know it takes only so many watts to make the motor operate. The watts are found by multiplying the amperes times the volts. In this case, 115 times 26 equals 2990 watts. If 230 is multiplied times 13 it too will equal 2990 watts. Note here that the 1 horsepower normally calls for 746 watts. If this is a 2-horsepower motor, it should draw 1492 watts from the AC line. What accounts for the extra wattage consumed? "SFA" is left blank. "HZ 60" means that the frequency of the line voltage should be 60 Hertz. "AMB 40" means that the temperature of the case or frame of the motor can be 40°C over the surrounding air temperature. Frame number ("FR") is shown as TP145T. Code is K. The NEMA letter designations following the frame number mean:

Figure 8–3. Westinghouse nameplate.

C—face mount.

H—has 2F dimension larger than same frame without H suffix.

J—face mount to fit jet pumps.

K—has hub for sump-pump mounting.

M—flange mount for oil burner, 5½-inch rabbet diameter.

N—flange mount for oil burner, 6⅜-inch rabbet diameter.

T,TS—integral HP motor dimension standards set by NEMA in 1964.

Y—nonstandard mounting: see manufacturer's drawing.

Z—nonstandard shaft extension (N, U dim.)

"INS" refers to the insulation related in terms of A, B, F, H. See the index for the temperatures associated with each letter. "TIME" can be "CONT" for continuous or "INT" for intermittent operation. "HSG OPEN" means the housing or frame has an opening for air to flow through. Also note that the nameplate has the wiring diagram for connecting the motor to run either clockwise or counterclockwise. *To reverse, interchange red and black leads.* The nameplate also tells you that the motor is permanently lubricated at the factory so no relubrication is required.

Troubles

Now that you know what the nameplate symbols mean, take a closer look at what may happen in the way of trouble if you do not follow the parameters indicated by the nameplate. In most instances where there is motor trouble, one or more of the nameplate specs has been ignored.

1. Do not operate motors at other than ±10 percent of the nameplate voltage.
2. Do not operate motors on nominal power source frequencies that are other than that indicated on the nameplate.
3. Do not overload the motor in excess of nameplate output rating.
4. Do not exceed temperature of nameplate insulation class.
5. Do not change the value of capacitance indiscriminately.
6. Do not subject the motor to duty cycles for which it was not designed.

This last paragraph applies mainly to permanent split-capacitor motors. Motor-start capacitors, used with split-phase motors, are usually specified to achieve maximum starting torque and/or minimum locked current, and deviations are not usually made by the user. Deviating to a higher value of capacitance will provide increases in starting torque and, in some cases, speed, but at the risk of certain hazards such as higher winding temperatures, shorter motor life, nuisance operation of the overload protectors, and increases in the level of noise and vibration.

Table 8–1 indicates the problems caused by variations in the nameplate parameters of an electric motor. Keep in mind that a motor is designed for certain conditions. If these conditions are exceeded, you can expect trouble. This may have been the case where you have been called to look at a motor that has been abused. If you repair it and then reinstall it in the same conditions, the same thing may happen again. Investigate the operational conditions for the motor before replacing it.

Bearings

No matter which type of special-application motor you may choose, it will have bearings. The type of bearings it uses has much to do with maintenance and noise level. Ball bearings, for instance, are

Table 8–1. Performance Parameters Adversely Affected by Nameplate Deviations

Nameplate Parameters	Torque	Speed	Temperature	Noise	Vibration	Thermal O/L Protectors	Current Sensitive	Centrifugal Cut-Outs	Capacitor Life	Motor Life
Voltage	X	X	X	X	X	X	X		X	X
Frequency	X	X	X	X	X	X	X	X	X	X
Horsepower (Torque)		X	X	X	X	X	X			X
Temperature	X	X				X	X			X
Capacitor	X	X	X	X	X	X			X	X
Duty			X			X	X	X	X	X

(Courtesy Bodine Electric Co)

rugged, precision-made, and noisier than sleeve bearings. The ball bearing is called for when heavy axial and radial thrust loads are encountered. The rugged, single-shielded ball bearings used in motors are usually oversized and can withstand the thrusts and shocks encountered in heavy-duty industrial applications. Bearings are preloaded with spring washers to take end play out of the bearing and contribute to quieter operation (Fig. 11–14). The open ball bearings shown in Fig. 8–4 are repackable; that is, they can be repacked with grease if necessary. The covered, or enclosed, type also shown in the illustration is not easily repacked.

Large babbitt-lined sleeve bearings are precision-machined to extremely close tolerances for accurate alignment, quiet operation, and long, dependable operation (Fig. 8–5). Accuracy of alignment helps ensure that the shaft is supported by a maximum bearing surface, thus reducing bearing wear to a minimum.

Wide-angle bearing oil grooves distribute the lubricant evenly over the entire bearing surface. This eliminates the possibility of "dry spots" and metal contact. This even, continuous distribution of freshly filtered oil cleans as it lubricates. That means it is capable of getting rid of any abrasive foreign particles.

Three-Speed Motors

The permanent-split capacitor motor can be easily changed in speed by simply flicking a switch from *Lo* to *Med* to *Hi* speed. Fig. 8–6 shows how the three windings are connected to allow for operation on three different speeds. Note that the three windings are all in the cir-

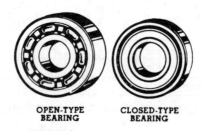

OPEN-TYPE
BEARING

CLOSED-TYPE
BEARING

Figure 8–4. Motor ball bearings.

Figure 8–5. Motor sleeve bearings.

cuit when the *Lo* speed is selected. More windings mean slower motor speed.

These split capacitor types are usually found on fans and other devices which need very little horsepower. In fact, take a look at Table 8–2 to see how the millihorsepower compares with horsepower in fractional form. *Milli* means $\frac{1}{1000}$ of a horsepower. Therefore, we can see that a $\frac{1}{20}$-horsepower motor is about 50 millihorsepower in size. Today there is a demand for conversion to metric. If you care to convert to the metric method of indicating power rating of a motor, just use the term *watts*. A 1-horsepower motor will draw 746 watts from the power source. That means that a 1-horsepower motor will be rated at 746 watts metric. A $\frac{1}{1000}$-horsepower motor will be rated as 0.746 watt in metric.

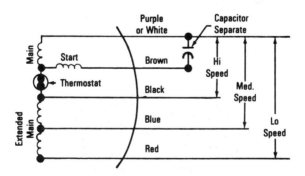

Note: Some motors will have Blue as Hi, Orange as Med. and Green as Lo.

Figure 8–6. Color-coded wires show the relationship to speed of the multispeed motor.

Table 8–2. Motor Power Output Comparison

Watts Output	*MHP	**HP
.746	1	1/1000
1.492	2	1/500
2.94	4	1/250
4.48	6	1/170
5.97	8	1/125
7.46	10	1/100
9.33	12.5	1/80
10.68	14.3	1/70
11.19	15	1/65
11.94	16	1/60
14.92	20	1/50
18.65	25	1/40
22.38	30	1/35
24.90	33	1/30
26.11	35	
20.80	40	1/25
37.30	50	1/20
49.70		1/15
60.17		1/12
74.60		1/10

*Millihorsepower
**Fractional horsepower
NOTE: Watts output is the driving force of the motor as calculated by the formula:

$$\frac{TN}{112.7} = WO$$

where T = torque (oz. ft.), N = speed (rpm).

Fig. 8–7 shows a permanent-split capacitor motor used for an air-conditioner blower. It is made in ⅙ to ½ horsepower for use on 115 or 230 volts of 60 Hz. It has an automatically-operated thermal protector built in. Note the location of the capacitor on top of the motor. It is not necessary to locate it on the motor. It is possible to place the capacitor somewhere in the air-conditioner cabinet and connect it to the capacitor by longer leads. It, too, is multispeed in design.

In Fig. 8–8 you see a ¼-horsepower, two-speed, yoke-mounted fan

Figure 8–7. Permanent-split capacitor fan motor.

motor. It, too, is a permanent-split capacitor motor. This one is totally enclosed and has a 1100- and 800-rpm speed capability. It was especially made for column-mounted air circulators. It relies on air passing over the motor for cooling. Since it is designed for use on fans, it is only proper to expect it to operate normally under those conditions. If it is used elsewhere, the limitations should be noted and taken into account when it comes to proper ventilation. Note the pull-chain-operated switch to turn it on or change the speed.

Another split capacitor blower motor is shown in Fig. 8–9. It has the capability for four speeds. Take a look at the speeds and volts as well as the horsepower shown in the illustration. It is designed for direct-drive furnace blowers and other air-over fans. Speed is selected by making the proper connection when the motor is hooked up to the furnace. Instructions that come with the motor tell which color wire produces which speed.

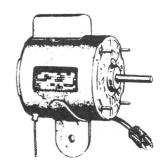

Figure 8–8. A $\frac{1}{4}$-horsepower, two-speed, yoke-mounted, enclosed motor.

Figure 8–9. Direct-drive furnace blower motor with leads. This is a three-speed motor that pulls up to 6.2 amps on 115 volts, 60 Hz. Thermally protected, the motor has a Frame 42 design. *(Courtesy Westinghouse)*

Instant-Reversing Motors

Reversing a motor can be rather hard on the motor and component parts. It takes a great deal of effort to stop the motor from its top speed and then cause it to run in the opposite direction. The bearings take quite a beating, as do the start switch and start capacitor. Some motors have a built-in relay for reversing, and others have to rely on a reversing switch (Fig. 8–10).

Drum switches are manually-operated three-position, three-pole switches that carry a horsepower rating and are used for manually reversing single- or three-phase motors. They are available in several sizes and can be spring-return-to-off (momentary contact) or maintained contact. Separate overload protection, by manual or magnetic starters, must usually be provided since drum switches do not include this feature.

Figure 8–10. Drum switch.
(Courtesy Square D)

Reversing Starter

Reversing the direction of motor shaft rotation is often required. Three-phase squirrel-cage motors can be reversed by reconnecting any two of the three line connections to the motor. By interwiring two contactors, an electromagnetic method of making the reconnection can be obtained.

Fig. 8–11 shows the contacts (*F*) of the forward contactor, when closed, connect lines 1, 2, and 3 to motor terminals *T1*, *T2*, and *T3*, respectively. As long as the forward contacts are closed, mechanical and electrical interlocks prevent the reverse contactor from being energized.

When the forward contactor is deenergized, the second contactor can be picked up, closing its contacts (*R*) which reconnect the lines to the motor. Note that by running through the reverse contacts, line 1 is connected to motor terminal *T3*, and line 3 is connected to motor terminal *T1*. The motor will now run in the opposite direction.

Whether operating through either the forward or reverse connector, the power connections are run through an overload relay assembly, which provides motor overload protection. A magnetic reversing starter, therefore, consists of a starter and contactor, suitably interwired, with electrical and mechanical interlocking to prevent the coil of both units from being energized at the same time.

Applications

One of the many uses for an instantly reversing motor is shown in Fig. 8–12. The ⅓-horsepower motor shown here is used for parking lot gates. The motor has three leads; one is common and the other two are

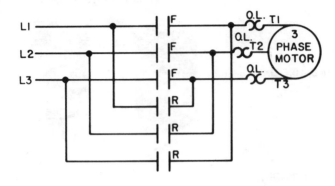

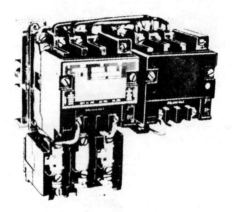

Figure 8–11. Reversing start schematic and starter. *(Courtesy Square D)*

used for reversing and forward. Simple toggle or individual single-pole relays are used to control direction of rotation.

Severe-Duty Motors

Severe-duty motors (Fig. 8–13) are used in abnormal conditions. Typical applications include dairy and food-processing equipment where hosing down of motors is routinely done, installations associated

Figure 8–12. An instant-reversing motor is needed for such jobs as parking gate operation. *(Courtesy Doerr)*

with plating equipment, mining equipment (not requiring explosion-proof construction), tropical applications, pumps or other devices located in unusually humid, wet, salty, acidic, alkaline, or dirty locations. This motor is not suitable for explosive and combustible-dust atmospheres. Use an explosion-proof type of motor for these applications.

The housing is rolled steel. A special sealing compound is used between the end shields and the motor stator. The nameplate is usually stainless steel; the bearings are usually ball and are lubricated for many years of service.

In some cases, such as in the process industries, the conditions for motor operation are so demanding that an aluminum motor has been

Figure 8–13. Severe-duty motor. This one is enclosed so that hosing down will not affect the interior.

designed (Fig. 8–14). These motors are totally enclosed, with tight rabbet fits between the frame and end shields. The windings, bearings, air gap, and other internal parts are protected from corrosion.

Foreign material is kept out by a durable neoprene slinger on the motor shaft which throws off water or any contaminant before it reaches the bearing. A gasket locks dust and moisture out of the conduit box. A lead seal gasket is used between the box and the frame to isolate completely the interior of the motor. Epoxy paint is used to coat the motor. This type of motor may be capacitor-start or three-phase. It comes in horsepower ratings from ¼ to ½ in capacitor-start, single-phase, and from ⅓ to 50 horsepower in 3ϕ. All single-phase motors are 1725 rpm, but the 3ϕ can be both 1725 and 3450 rpm.

Farm-Duty Motors

Heavy-duty motors designed specially for severe farm duty are capacitor-start induction-run types that furnish high starting torque with normal current. They are gasketed throughout for environmental

Figure 8–14. Special aluminum housing for a severe-duty motor.

protection and have double-sealed ball bearings with a water flinger on the shaft end to protect the motor and bearings from contaminants. An oversize conduit box makes wiring easy. Grounding provision is also included. Low-temperature thermal-overload protectors can be used, but a manual reset overload button (with rubber weather boot) is standard for maximum operator safety.

In Table 8–3 are shown some of the variations on the capacitor-start and three-phase motor available for farm use by only one of the many motor manufacturers.

The farm-duty motor is shown in Fig. 8–15. It is used for pumps, conveyors, and poultry equipment, and on other farm machinery. The motor shown in Fig. 8–16, however, is designed for use in direct- and belt-driven fans where the motor is located in the air stream, such as on the poultry-house and barn exhaust fans. They operate without trouble in lint, dust, and dirt. All of these have ball bearings since the noise level is not important. Two speeds and two voltages are available. Keep in mind that most farms have only single-phase AC and must rely on this type of motor to serve their needs.

Figure 8–15. Farm-duty motor with box mounted on the side and capacitor on top. *(Courtesy Leeson)*

Table 8–3. Single-Phase Capacitor-Start/Induction-Run Totally-Enclosed, Fan-Cooled 104°F(40°C) Ambient

HP	RPM	Frame	Voltage	Hertz	Overload Protector
Rigid Mount					
1/3	1725	E56	115/208-230	60	Manual
1/2	1725	D56	115/208-230	60	Manual
3/4	1725	E56	115/208-230	60	Manual
1	1725	F56	115/208-230	60	Manual
1 1/2	1725	H56H	115/208-230	60	Manual
2	1725	J56H	230	60	Manual
C-Face Round Body Without Base					
1/3	1725	E56C	115/208-230	60	Manual
1/2	1725	D56C	115/208-230	60	Manual
3/4	1725	E56C	115/208-230	60	Manual
1	1725	F56C	115/208-230	60	Manual
1 1/2	1725	H56C	115/208-230	60	Manual
2	1725	J56C	230	60	Manual
Three-Phase **Totally-Enclosed, Fan-Cooled, 104°F(40°C) Ambient**					
Rigid Mount					
1/3	1725/1425	C56	208-230/460	60/50	Manual
1/2	1725/1425	D56	208-230/460	60/50	Manual
3/4	1725/1425	D56	208-230/460	60/50	Manual
1	1725/1425	E56	208-230/460	60/50	Manual
1 1/2	1725	F56	208-230/460	60	Manual
2	1725	F56	208-230/460	60	Manual
C-Face Round Body Without Base					
1/3	1725	C56C	208-230/460	60/50	Manual
1/2	1725/1425	D56C	208-230/460	60/50	Manual
3/4	1725/1425	D56C	208-230/460	60/50	Manual
1	1725/1425	E56C	208-230/460	60/50	Manual
1 1/2	1725	F56C	208-230/460	60	Manual
2	1725	G56C	208-230/460	60	Manual

(Courtesy Leeson)

Figure 8–16. Belted fan and blower motor. *(Courtesy Leeson)*

Table 8–4 shows the variety of motors, speeds, and voltages available in the single-phase design for farm use.

High-Slip Motors

High-slip motors have high starting torque. They are designed for use where frequent or protracted starting under heavy loads is required. These could be used in hoists, cranes, punch presses, elevators, oilwell loads, shears, and bending brakes. The high—8 to 13 percent—running slip and varying-speed characteristic qualifies these motors for drives with high inertia. They are made in sizes from 3 to 7½ horsepower. This is well above our preset limit of 1 horsepower for the book. However, they are mentioned here to indicate that such a motor does exist and can be used for very exact purposes where other motors may be overloaded. They are made in 3ϕ only since they are intended for industrial applications. The voltages are $^{230}/_{460}$ at 60 Hz. Most have Class B insulation (Fig. 8–17).

Table 8–4. Farm-Duty Motors (Single-Phase)

HP	Speed (rpm)	Volts	NEMA Frame	Bearings	Therm. Prot.	Full-Load Amps
¼	1725	115	48	Ball	None	4.5
		115	48	Ball	Auto	4.5
	1140	115	56	Ball	None	5.3
		115	56	Ball	Auto	5.3
⅓	1725	115	48	Ball	None	5.3
		115	48	Ball	Auto	5.3
		230	48	Ball	None	2.7
		230	48	Ball	Auto	2.7
	1140	115	56	Ball	None	7.0
½	1725	115	56	Ball	None	8.0

Rotation—Shaft rotation easily reversed by electrical reconnection. *(Courtesy General Electric)*

Motor Slip

When the induction motor is started without a load, the magnetic field of the stator revolves past the rotor at synchronous speed. This produces a maximum current in the rotor conductors and causes the rotor to revolve in the direction of the rotating magnetic field. As the rotor speed increases, the relative speed of the moving magnetic field decreases, since the rotor tends to "catch up" with the rotating field. If the rotor speed becomes equal to the rotating magnetic field, there is no longer any relative motion between the rotor and the magnetic field. This means no cutting of lines of force and no induced current in the rotor. That, in turn, means that there is no torque or turning effort in the rotor. As you can see from this, the speed of the rotor is always less than that of the rotating field. That means a motor with 1800 rpm designed as the rotating field will have about 1725 rpm for the rotor speed, and a 3600-rpm motor really has a speed of 3450 rpm.

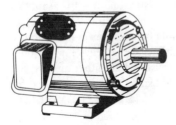

Figure 8–17. High-slip motor.

The percent of slip in an induction motor is defined as the difference between the synchronous speed and the actual speed to the synchronous speed. The 1725 rpm of the 1800-rpm motor calls for a 5-percent slip ratio. The same percentage is found for the 3450- or 3600-rpm motor. Therefore, you can see the 8- to 13-percent slip motor is really a high-slip or high-torque motor.

Explosion-Proof Motors

Fractional-horsepower explosion-proof motors are available in ¼ to 1 horsepower. They are made in split-phase and capacitor-start designs. They have sealed ball bearings and are totally enclosed, nonventilated types in the ¼-horsepower size. The ⅓- to 1-horsepower motors are totally enclosed and fan-cooled (Fig. 8–18).

Explosion-proof motors are made to meet Underwriters Laboratories Standards for use in hazardous (explosive) locations, and this is indicated on the nameplate. Certain locations are hazardous because the atmosphere does or may contain gas, vapor, or dust in explosive quantities. The *NEC* (*National Electrical Code*) divides these locations into classes and groups according to the type of explosive agent which may be present. Some of these classes and groups are listed below. For a complete list, refer to Article 500 of the latest edition of the *National Electrical Code*.

Class I (Gases, Vapors)

Group A   Acetylene. Motors are not available for this group.

Group B   Butadiene, ethylene oxide, hydrogen, propylene oxide. Motors are not available for this group.

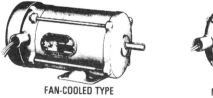
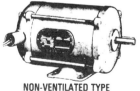

FAN-COOLED TYPE   NON-VENTILATED TYPE

Figure 8–18. Explosion-proof motors, vented and nonventilated types.

Group C Acetaldehyde, cyclopropane, diethyl ether, ethylene, isoprene.

Group D Acetone, acrylonitrile, ammonia, benzene, butane, ethylene dichloride, gasoline, hexane, methane, methanol, naphtha, propane, propylene, styrene, toluene, vinyl acetate, vinyl chloride.

Class II (Combustible Dusts)

Group E Aluminum, magnesium, and other metal dusts with similar characteristics.

Group F Carbon black, coke, or coal dust.

Group G Flour, starch, or grain dust.

Other Motors

Open, drip-proof motors are designed for use in areas that are reasonable dry, clean, well ventilated, and usually indoors. If installed outdoors, it is recommended that the motor be protected with a cover that does not restrict the flow of air to the motor.

Totally enclosed motors are suitable for use where exposed to dirt and moisture, and in most outdoor locations, but not for very moist or for hazardous (explosive vapor, dust-filled atmosphere) locations. Severe-duty enclosed motors are suited for use in corrosive or excessively moist locations.

Metric Motor Ratings

For some time the United States has been considering the idea of switching to metrics for measurements. However, the electrical field has always been, in most instances, metric. The volt, ampere, ohm, and other units of measurement are already used in metrics, too. Only a few terms in the electrical field are still English in their measurements. One of these is the horsepower. The metric unit for measuring power is the watt. It takes 746 watts to equal 1 horsepower. Therefore, the newer (SI) International Standard for measurement is the watt or kilowatt. It takes 1000 watts to equal one kilowatt. That means 746 watts equal 0.746 kilowatt. There are some standard motor sizes which can be quickly converted to the metric rating of kilowatt instead of horsepower. They follow below:

Table 8–5.   Relationship of Horsepower to Kilowatt for Motor Ratings

Horsepower	Kilowatt (kW)*
$\frac{1}{20}$	0.025
	0.035
	0.05
	0.071
$\frac{1}{8}$	0.1
$\frac{1}{6}$	0.14
$\frac{1}{4}$	0.2
$\frac{1}{3}$	0.28
$\frac{1}{2}$	0.4
1	0.8
$1\frac{1}{2}$	1.1
2	1.6
3	2.5
5	4.0
7.5	5.6
10	8.0

*James W. Polk, A *Preview of Metric Motors,* Westinghouse Electric Corporation.

PART THREE

Maintenance and Repair

CHAPTER 9

General Maintenance

Proper maintenance procedures mean the difference between a short and long motor life. Trouble-free operation is the main objective of the manufacturer and owner of a motor. If the motor does not operate properly, it means a loss of money or effort. Procedures are specified by the manufacturer to be followed for proper operation of electric motors. The procedures are, in most instances, very simple and easy to follow. Whenever these precautions are neglected, the obvious problems result. This chapter deals with these problems and how to remedy them once they have been identified.

One of the most often encountered troubles in electric motors is low voltage. Low voltage can be caused by the wrong wire size in the circuit. Table 9–1 shows the circuit wire sizes for individual single-phase motors. Note how the length of the run from the distribution panel to the motor affects the size of wire to be used. Selecting the wrong size wire can cause a voltage drop along the wire that will result in the motor receiving the lower voltage. As a result of low voltage, the motor will attempt to draw more current. As the current increases, the heat generated increases by the square of the current. This further increases the voltage drop along the wire and, in turn, again reduces the line voltage at the motor.

In general, motors should be checked for a number of wear indica-

Table 9–1. Circuit Wire Sizes for Individual Single-Phase Motors

Horsepower of Motor	Volts	Approximate Starting Current Amperes	Approximate Full-Load Current Amperes	Feet	Length of Run in Feet (from Main Switch to Motor)							
					25	50	75	100	150	200	300	400
1/4	120	20	5	Wire Size	14	14	14	12	10	10	8	6
1/3	120	20	5.5	Wire Size	14	14	14	12	10	8	6	6
1/2	120	22	7	Wire Size	14	14	12	12	10	8	6	6
3/4	120	28	9.5	Wire Size	14	12	12	10	8	6	4	4
1/4	240	10	2.5	Wire Size	14	14	14	14	14	14	12	12
1/3	240	10	3	Wire Size	14	14	14	14	14	14	12	10
1/2	240	11	3.5	Wire Size	14	14	14	14	14	12	12	10
3/4	240	14	4.7	Wire Size	14	14	14	14	14	12	10	10
1	240	16	5.5	Wire Size	14	14	14	14	14	12	10	10
1 1/2	240	22	7.6	Wire Size	14	14	14	14	12	10	8	8
2	240	30	10	Wire Size	14	14	14	12	10	10	8	6
3	240	42	14	Wire Size	14	12	12	12	10	8	6	6
5	240	69	23	Wire Size	10	10	10	8	8	6	4	4
7 1/2	240	100	34	Wire Size	8	8	8	8	6	4	2	2
10	240	130	43	Wire Size	6	6	6	6	4	4	2	1

tors at least twice a year. In the case where heavy usage is encountered, the motor should be checked more often. A maintenance schedule can be set up according to previous experience with the various types of motors. Manufacturer's recommendations should be followed closely for best operation with the least amount of trouble.

Each type of small motor is covered in this chapter for its maintenance and troubleshooting procedures. The split-phase, the capacitor-start, the permanent-split capacitor, the shaded-pole, the three-phase, and the brush-type (series, permanent-magnet, shunt, and compound) motors are covered in detail to aid you in locating the possible source of trouble and in maintaining the motors so fewer troubles will appear during normal operation.

General Troubleshooting Procedures

Before servicing or working on equipment, disconnect the power source. This applies especially when servicing equipment using thermally protected automatic restart devices instead of manual restart devices and when examining or replacing brushes on brush-type motors.

Clean the motor environment regularly to prevent dirt and dust from interfering with the ventilation or clogging the moving parts.

Capacitor-Equipped Motors

The capacitors used for capacitor-start and capacitor-start, capacitor-run motors are dangerous when charged. Before servicing motors with capacitors, always discharge the capacitor by placing a conductor (a screwdriver, usually) across its terminals before you touch the terminals with any part of your body.

In many cases, easy-to-detect symptoms will indicate exactly what is wrong with your fractional-horsepower motor. However, since general types of motor trouble have similar symptoms, it is necessary to check each possible cause separately. The tables provided in this chapter list some of the more common ailments of small motors, with suggestions as to probable causes.

Most common motor troubles can be checked by some test or inspection. While the order of these tests rests with the troubleshooter, it is advisable to make the easy ones first. In diagnosing troubles, a combination of symptoms will often give a definite clue to the source of the trouble. For example, if a motor will start but heating occurs, there is a good likelihood that a short or ground exists in one of the windings.

Centrifugal starting switches, found on many types of fractional-horsepower motors, are occasionally the source of motor trouble. Such switches have a finite life and they wear in many ways, depending on the design and usage. If the switch sticks in the open position, the motor will not start. When stuck in the closed position, the motor will normally operate at a slightly reduced speed and the start winding will quickly overheat. The motor may also fail to start if the contact points of the switch are out of adjustment or coated with oxide. It is important to remember, however, that any adjustment of the switch or contacts should be made only at the factory or by an authorized service center.

Brush-Type Motors

Because of wear, brushes and commutators on commutated motors require more maintenance than nonbrush types. The wear rate of brushes depends on many parameters (armature speed, amperage conducted, duty cycle, and humidity, to name a few). For optimum performance, brush-type motors need periodic user-maintenance. The maintenance interval is best determined by the user. Inspect the brushes regularly for wear (replace in same axial position). Replace brushes when their length is less than ¼ inch (7mm). Periodically remove carbon dust from the commutator and from inside the motor. This can be accomplished by occasionally wiping them with a clean, dry, nonlinting cloth. Do not use lubricants or solvents on the commutator. If necessary, use No. 0000 or finer sandpaper to dress the commutator. However, if pitted spots still appear, the commutator should be reground by an experienced electric-motor service center.

Maintenance of the Split-Phase Motor

The split-phase motor will perform well with little attention as long as it is not overloaded. For a motor that performs efficiently and with little trouble, there are a few maintenance suggestions.

1. Inspect the motor at least twice a year. Check for obvious wear. With a belt-driven arrangement, the belts may be too tight and may cause uneven wearing of the two bearings by placing undue pressure on them. Furthermore, as the bearings wear down, the rotor can strike the stator poles and permanently damage the rotor. This can distort the stator windings and cause them to become damaged severely enough to require rewinding (Fig. 9–1).

2. Check the centrifugal switch (Fig. 9–2). It should shut off before the motor has come to final running speed, at 75 percent of the rated speed. This check can be made with some degree of precision by using a tachometer to test the operating speed and the click-out speed of the switch. On the 1800-rpm motor, the switch should click out at approximately 1350 rpm. On the 3600-rpm motor, the start switch should click out at 2700 rpm. The nameplate speeds will be 1725 rpm for the 1800-rpm motor and 3450 rpm for the 3600-rpm motor. The switch should click in just before

RESET CIRCUIT BREAKER

Figure 9–1. Stator windings and circuit breaker of a motor. This stator should be rewound.

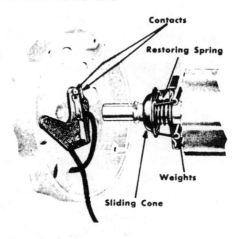

Figure 9–2. **Centrifugal switch.** *(Courtesy Bodine)*

the motor coasts to a dead stop after it has been disconnected from the power line.

Do not overlook the screws that hold the centrifugal switch in place when the motor is disassembled for inspection or maintenance. Make sure the switch screws have not become loose. Since the switch is used each time the motor is turned on or off, check the mounting screws when cleaning the points of the switch and tighten the screws to make sure. See Fig. 9–3 for a typical centrifugal switch.

3. Check for overloads. Sometimes a load will build up because of increased friction and wear throughout the driven system. Check the motor temperature for any indication that it is running hot. Normal motor temperature is one where you can place your hand on the motor and it will feel warm. A hot temperature is hot to the touch after a few seconds. Make sure you check the load again before continuing to use the motor under these conditions.

Overload fuses, or cut-outs, should be part of the motor design. If they are not, make sure a fuse or cut-out is inserted in series with the line that is supplying power to the motor. If the motor's reset button keeps cutting out and turning the motor off, make sure you find the cause of the overload and correct it before the motor becomes shorted and the windings are burned. Some mo-

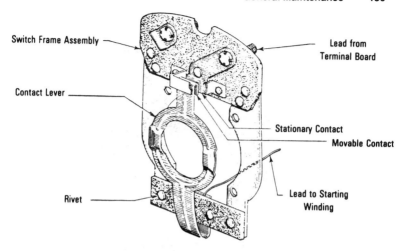

Figure 9–3. Centrifugal switch.

tors have a relay to open the auxiliary coil circuit (Fig. 9–4). When power is applied to the line, current flows through the heavy main winding. This closes the contacts connecting the start winding to the supply line voltage (Fig. 9–5). With the pickup in speed, the current is reduced. This reduction in current causes the current-sensitive relay solenoid to deenergize (drop out), which causes the contacts to open, removing the auxiliary (start) winding from the circuit (see schematic in Fig. 9–6).

4. Lubricate the motor. This is especially true of the sleeve-type motor bearing. A motor that is run continuously requires more attention than one that is run occasionally. However, it is easy to forget to lubricate a motor. A typical example would be the split-phase fan motor that is mounted inside the furnace cabinet in hot-air furnaces. As the saying goes, "Out of sight, out of mind." Lubrication for this motor is usually forgotten.

   The motor should be lubricated to the manufacturer's specifications. Provide enough oil, but do not overdo it. Too much oil collects dust and creates problems of another sort.

5. Check the centrifugal switch. The sliding member of the switch assembly should move freely on the motor shaft. If the contact points are pitted or burned, use a fine-grade sandpaper and polish

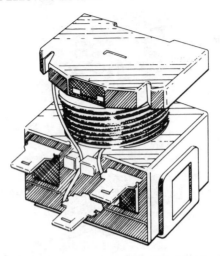

Figure 9–4. Current-sensitive relay used to take the start winding out of the circuit once the motor has started and come up to 75 percent of synchronous speed.

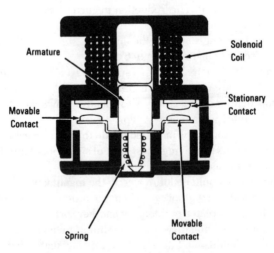

Figure 9–5. Cutaway view of the current relay.

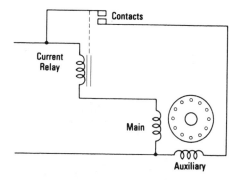

Figure 9–6. Schematic of the current relay in the motor circuit.

them so they seat against one another properly. Do not use metallic abrasive papers; they will leave particles of metal that can cause excessive wear if they get into the bearings, or even shorts if they stick to parts of the switch.

6. Check the alignment and pulley mechanism. Watch for an out-of-alignment condition where the shaft is not matched with the load. Sometimes the motor is damaged by an increased load that is caused by misalignment of the drive shaft and the driven load. Check to make sure the pulley on the motor and the pulley on the load are aligned and that the belt is not twisted in any way (Fig. 9–7).

7. Check the wiring when installing a motor for the first time or when moving a motor to a different load condition. Make sure the rating of the motor and the wires match. If the motor draws more current than the wires can handle safely, there may be a drop in the voltage reaching the motor. This means the motor is operating under reduced voltage conditions and an increase in heat generated by the motor.

A common problem with saw motors and portable motor-driven tools is that the extension line is too small to handle the current drawn by the motor when started. Too much voltage is dropped along the line and the available voltage for the motor is severely limited. This often happens in on-the-job building situations where carpenters use extension cords from one newly built house to another when the power is not yet available to the new house under construction.

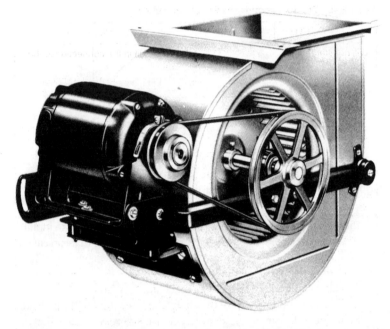

Figure 9–7. Split-phase motor driving a home hot-air furnace blower. Note the belt drive. Also note the base adjustments so the motor can be aligned properly to prevent excessive belt wear or motor bearing damage. *(Courtesy Lennox)*

Identification of Split-Phase Motor

In a classroom situation you may be given a motor that has been pretty well stripped down and is hard to identify. Just remember that the split-phase motor will have a start winding made of small-diameter wire and a run winding made of a larger wire. The capacitor-start motor has windings with wires of the same diameter.

Troubleshooting

Table 9–2 is a quick reference to troubleshooting a split-phase motor.

Table 9–2. Troubleshooting the Split-Phase Motor

Trouble	Probable Cause
Will not start at all.	Open circuit in the connection to the line. Open circuit in the motor winding. Contacts of the centrifugal switch are not closed. Start winding open.
Will not always start, even with no load, but will run in either direction when started manually. Starts, but heats rapidly.	Contacts of centrifugal switch are not closed. Start winding open. Centrifugal switch is not opening. Winding short-circuited or grounded.
Starts, but runs too hot.	Winding short-circuited or grounded.
Wll not start, but will run in either direction when started manually; overheats.	Contacts of centrifugal switch are not closed. Start winding open. Winding short-circuited or grounded.
Motor gets too hot; reduction in power evident.	Winding short-circuited or grounded. Sticky or tight bearings. Interference between stationary and rotating members.
Motor blows fuse, or will not stop when it is turned to the OFF position.	Winding short-circuited or grounded. Grounded near the switch end of the winding.

Maintenance of Capacitor Motors

Capacitor motors (Table 9–3), like the split-phase types, will operate for years will little or no maintenance, but can develop some troubles with the capacitors and the start switches. Inspect the motor twice a year. Check for wear that may be obvious to the eye. Belts may be too tight and cause wear on the bearings if they are the sleeve type. The bearings may be damaged by prolonged use with a belt drive that is too tight.

Check the centrifugal switch. It should click out before the motor has reached full speed. The ideal place for this to occur is at 75 percent of rated speed. This can be checked with some precision if a tachome-

Table 9–3. Troubleshooting the Capacitor-Start Motor

Trouble	Probable Cause
Will not start.	Open circuit in connection to the line. Open circuit in motor winding. Contacts of centrifugal switch are not closed. Defective capacitor. Start winding open.
Will not always start, even with no load, but will run in either direction when started manually.	Contacts of centrifugal switch are not closed. Defective capacitor. Start winding open.
Starts, but heats rapidly.	Centrifugal switch is not opening. Winding is short-circuited or grounded.
Starts, but runs too hot.	Winding is short-circuited or grounded.
Will not start, but will run in either direction when started manually; overheats.	Contacts of the centrifugal switch are not closed. Defective capacitor. Start winding open. Winding short-circuited or grounded.
Reduction in power; motor runs too hot.	Winding short-circuited or grounded. Sticky or tight bearings. Interference between the stationary and rotating members.
Motor blows fuse or will not stop when the switch is turned to OFF position.	Winding short-circuited or grounded. Grounded near the switch end of the winding.

ter is used to obtain the running speed. For example, the 1800-rpm motor should have the start switch click out at 1350 rpm. The nameplate speeds will be 1725 rpm for the 1800-rpm motors and 3450 rpm for the 3600-rpm motors. The switch should click in just before the motor coasts to a dead stop after it has been disconnected from the power line.

Check for overloads. Sometimes a load will build up due to increased friction and wear throughout the driven system. Watch the motor temperature for any indication that it is running hot. This means, in most cases, that if you place your hand on the motor and have to pull it back quickly because the motor is too hot.

Make sure the overload protection is working properly. A fuse or circuit breaker of the proper size should be in the circuit to protect it in case of an overload.

Check the centrifugal switch to see if the sliding part is moving freely. If the contact points are pitted, clean them with a piece of sandpaper. Do not separate the points too much or they may not serve their purpose properly.

Do not overlook the screws that hold the centrifugal switch in place when you have the motor disassembled for service or to look for damage. In some instances, be sure that the switch screws have not worked loose. Since the switch is used each time the motor is turned on or off, it is best to check when you clean the points of the switch. Just tighten the screws to make sure.

If pulleys are used, check their alignment to be sure that the belt is not out of line. This causes undue wear on the bearings of the motor and the driven device.

Check the wiring when you install a motor for the first time. It is a good idea to check the wire size if you are changing the location of a motor (Fig. 9–8). If the motor draws too much current, the wire will cause a voltage drop, which means there is less voltage than the motor specifications call for.

Keep in mind that the starting current of the capacitor motor is more than the running current. Check it out and make sure the protective device (fuse or circuit breaker) is capable of handling the start current for the short period without failing (Fig. 9–9).

Electrolytic Capacitors

The device that gives the capacitor-start motor its ability to start under load is the electrolytic capacitor. The electrolytic capacitor comes in a black case in most instances. This Bakelite case has two snap connectors which connect to the leads from the motor start windings. Either lead can go to either terminal on the capacitor since it is an AC electrolytic. No polarity is needed. However, it is necessary to obtain the right size of electrolytic for the motor. The correct microfarads and the correct working voltage should be obtained for best utilization of the motor's design qualities.

Table 5–1 will help you to match the ability of the electrolytic to the motor. There are three voltage ratings, so the correct voltage will

Figure 9–8. Capacitor-start motor used on a table saw. Note the size of the flexible cord used as a power cord.

Figure 9–9. Magnetic start switch for a 1-horsepower motor such as shown in Fig. 9–8.

have to be selected to make sure the capacitor is not damaged when connected into the start circuit.

It is a good rule to make sure you use the identical size capacitor as came out of the motor. If you cannot find the markings on the capacitor, look at Table 5–1 and select one that matches the characteristics of the motor as shown on its nameplate.

The Permanent-Split Capacitor Motor

This type of motor has a capacitor in the circuit even during the run period of the motor, a capacitor inserted in series with one of the two motor windings. The main winding is in parallel with the series coil-capacitor winding arrangement. The capacitor, which is somewhat expensive and bulky, is not taken out of the circuit when the motor has started. This is not a practical motor for heavy-duty starting. The troubles encountered with this type of motor are similar to those of the capacitor-start motor, with a few exceptions. Table 9–4 will indicate some of the more common problems and their probable causes.

The shaded-pole motor has some very interesting characteristics. One of them is the ability to start and run without two coils. The simple shaded-pole motor used for clocks and small fans is a one-coil device. It has a winding that is connected across the line at all times. It relies upon its core windings and the iron in the laminations to prevent overheating. Other types of shaded-pole motors may have four or six windings. These are closer in appearance to the regular split-phase-type motor. However, shaded-pole motors do not have start windings as such. They do have a large piece of metal (usually coat-hanger-size wire) that shorts a portion of the pole of each coil. This will identify the motor for troubleshooting purposes. Table 9–5 indicates the probable causes of shaded-pole motor troubles.

To repair, just remedy the cause. Remove grounds or shorts, or replace the sticky bearings, or make the proper clearance between rotating members and the stationary members of the motor. A drop of oil is all that is needed in some cases to ease the sticky or tight bearing.

Maintenance of the Three-Phase Motor

In order to completely check the three-phase motor, it is necessary to take it apart and look at each component part. Both inside and out-

Table 9-4. Troubleshooting the Permanent-Split Capacitor Motor

Trouble	Probable Cause
Will not start at all.	Open circuit in connection to line. Blown fuses; overload protector tripped or faulty. Open circuit in motor winding. Defective capacitor. Start winding open. Overloaded motor. Winding short-circuited or grounded. One or more windings open. Tight or seized bearings. Interference between stationary and rotating member. Wrong connection to motor. Improper or low voltage.
Will not always start, even with no load, but will run in either direction when started manually.	Defective capacitor. Start winding open. One or more windings open. Wrong connection to motor.
Starts, but heats rapidly.	Defective capacitor. Overloaded motor. Winding short-circuited or grounded. Tight or seized bearings. Interference between stationary and rotating member. Wrong connection to motor.
Runs too hot after extended operation.	Overloaded motor. Tight or seized bearings. Failure of ventilation (blocked or obstructed ventilation openings). Improper or low voltage from line. Worn bearings. High ambient temperature.
Excessive noise (mechanical).	Interference between stationary and rotating member. Worn bearings. Unbalanced rotor or armature (vibration). Poor alignment between motor and load; loose motor mounting. Amplified motor noises.

Table 9–4. (cont.)

Trouble	Probable Cause
Reduction in power; motor gets too hot.	Defective capacitor. Winding short-circuited or grounded. Tight or seized bearings. Interference between rotating and stationary member. Wrong connection of motor. Improper or low line voltage.

side should be checked. In Fig. 9–10 you see a rather old three-phase motor, still operational after about 40 years. Its basic design is the same as those made today. If we look closely at it, we will see it is ½ horsepower and was made by Peerless Electric Products.

There are four bolts holding the motor together. Remove these to take the end bells off. However, before going farther, look at the type of terminals used to connect power into the motor. The screw-on type, where the wire is pushed through a hole in the terminal and then the

Figure 9–10. A ½-horsepower three-phase motor.

Table 9–5.   Troubleshooting the Shaded-Pole Motor

Trouble	Probable Cause
Motor will not start at all.	Open connection to the line. Open circuit in motor winding. Overloaded motor. Winding short-circuited or grounded. Tight or seized bearings. Interference between stationary and rotating member. Wrong connection to motor. Improper line voltage. Low line voltage.
Starts, but heats rapidly.	Overloaded motor. Winding short-circuited or grounded. Tight or seized bearings. Interference between stationary and rotating member. Wrong connection to motor.
Runs too hot after extended operation.	Overloaded motor. Tight or seized bearings. Failure of ventilation. Improper or low line voltage. Worn bearings. High ambient temperatures.
Excessive noise (mechanical).	Interference between stationary and rotating member. Worn bearings. Unbalanced rotor (vibration). Poor alignment between motor and load or loose motor mounting. Amplified motor noises.
Reduction in power; motor gets too hot.	Winding short-circuited or grounded. Tight or seized bearings. Interference between stationary and rotating member. Wrong connection to motor. Improper or low line voltage.

cap is screwed down to hold it in place, is no longer used. Another thing to look at while it is still all together is the way it is lubricated. This is a sleeve-bearing-type motor—it has two places to put oil.

Once you remove the end bell, you will be able to see the rotor with its fan (Fig. 9–11). Do not lose any of the thin washers that may be on the shaft. These washers determine end play and can make a difference in the fan blades hitting the windings or not.

In Fig. 9–12 you can see the windings. They should be checked for damage resulting from the fan scraping them. In this view you can see that the insulation paper has deteriorated somewhat, but performance is not affected. Check the poles inside the motor to see if the rotor may have been touching them at any point. There would probably be a shiny spot, which could indicate that the bearings need replacing.

Fig. 9–13 shows what the motor looks like with both end bells removed. It is now possible to inspect both ends of the motor for possible motor winding damage. Fig. 9–13 shows some possible insulation damage where it extended past the windings. This may have been done when the motor was disassembled previously and the fan pushed back a little too far. It does not cause any problem with operation of the motor, however.

Thin washer for end play adjustment

Figure 9–11. **Motor with one end bell removed showing the fan that is part of the rotor assembly.**

Figure 9–12. Stator windings on a three-phase motor.

Figure 9–13. Three-phase motor with both end bells and rotor removed.

Fig. 9–14 shows a close-up view of the inside of an end bell. Note that the bearing is sleeve-type. Check for indications of wear and lack of lubrication.

In Fig. 9–15 you can get a better idea of what the rotor of a three-phase motor looks like. Note that it does not have a centrifugal switch; the switch is not needed in a three-phase motor. However, be sure not to lose any of the thin washers on the shaft. Also make a note of the location and the number of washers on each end. Fig. 9–16 shows the balancing weights placed in the fan to make the motor run smoothly.

Note the two oil cups in the end bell (Fig. 9–10). One is for the top of the bearing and the other is for making sure the lower half of the bearing is properly lubricated. In most cases you will find only one oil cup or one grease fitting per bearing. Better methods of distributing the oil have been devised.

General Maintenance

Maintenance of the three-phase motor is rather simple. Oil it if it has sleeve bearings. Make sure it is not overloaded if it has ball bear-

Figure 9–14.   Inside of the end bell.

Figure 9–15.   Three-phase motor rotor.

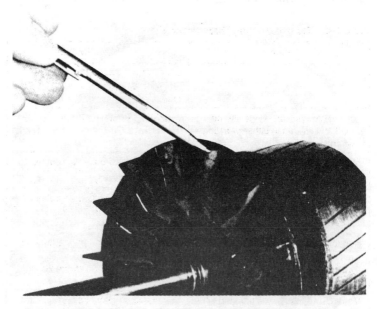

Figure 9–16.   Balancing weights on the rotor fan.

ings. It is almost trouble-free if properly loaded and operated on the correct line voltages. If it has flow-through ventilation, make sure the intakes are kept clean of dust, dirt, and the collection of fuzz.

Repair

It will become necessary to replace the bearings when they are worn—whether they are sleeve or ball types. It is a simple matter to replace either, so no special instructions are needed; just make sure they are replaced with the proper size. In some cases where the windings are burned or the insulation has been removed by overheating, it will be necessary to rewind the windings.

As you can see from Fig. 9–16, the rotor does not need any rewinding, since it has no windings. Just make sure the spacer washers are in the correct thickness and *number* when you reassemble the motor. The proper number refers to the number of washers needed to make sure there isn't too much end play in the rotor. Table 9–6 shows how to troubleshoot the three-phase motor.

Table 9–6. Troubleshooting Three-Phase Motors

Trouble	Probable Cause
Will not start.	Open circuit in connection line. Open circuit in motor winding. One or more windings open.
Will not always start, even with no load, but will run in either direction when started manually.	One or more windings open.
Starts, but heats rapidly.	Winding short-circuited or grounded.
Starts, but runs too hot.	Winding short-circuited or grounded.
Will not start, but will run in either direction when started manually; overheats.	Winding short-circuited or grounded. One or more windings open.
Reduction in power; motor gets too hot.	Winding short-circuited or grounded. Sticky or tight bearings. Interference between stationary and rotating members.
Motor blows fuse or will not stop when switch is turned to OFF position.	Winding short-circuited or grounded. Grounded near switch end of winding.

Maintenance of Brush-Type Motors

This heading refers to motors that can be used on AC or DC and are most commonly referred to as the universal type. This also includes the permanent-magnet, shunt, and compound DC motors. Keep in mind that the speed of a motor will also increase its life expectancy. A rough rule of thumb is that when the speed is reduced by 50 percent, the brush life is tripled. Table 9–7 shows the symptom, the probable cause, and the possible remedy for the small brush-type motor. In the previous tables the remedy was rather obvious. In this type of motor the remedy may not be so obvious, so the *Remedy* column has been added for your convenience.

To remedy, take the obvious steps. Lubricate if the bearings are sticking. Check the clearance to make sure the rotating and stationary members are not touching by checking bearings and proper armature for this motor. Check for shorts and remove such if possible; if not, rewind the motor. If line is open, complete the circuit in an approved manner meeting the *National Electrical Code* recommendations for that motor.

Troubleshooting the universal, series, shunt, and the compound DC motors is slightly different since the permanent-magnet type does not have a field winding. This is covered in Table 9–8.

Table 9–7. Troubleshooting the Permanent-Magnet Motor

Symptom	Probable Cause	Remedy
Blows circuit breaker and shaft turns very hard.	Bad bearing.	Replace the bearing. Also check the armature.
Blows circuit breaker and armature heats up.	Check for machine overload.	Replace armature and possibly rear-end bell assembly.
Does not run.	Worn brushes or brush hang-up	Replace or release brushes and check the commutator.
Does not run or runs too fast.	Open armature connection.	Replace the armature.
Motor runs in the wrong direction.	Reversed motor connections.	Reverse the polarity of the power supply.

Table 9–8. **Troubleshooting Universal, Series, Shunt, or Compound DC Motors**

Trouble	Probable Cause
Will not start.	Open circuit in connection to the line. Open circuit in motor winding. Worn brushes and/or annealed brush springs. Open circuit or short circuit in the armature winding.
Starts, but heats rapidly.	Winding short-circuited or grounded.
Starts, but runs too hot.	Winding short-circuited or grounded.
Sluggish; sparks severely at the brushes.	High mica between commutator bars. Dirty commutator or commutator is out of round. Worn brushes or annealed brush springs. Open circuit or short circuit in the armature winding. Oil-soaked brushes.
Abnormally high speed; sparks severely at the brushes.	Open circuit in the shunt winding.
Reduction in power; motor gets too hot.	Open circuit or short circuit in the armature windings. Sticky or tight bearings. Interference between the stationary and rotating members.
Motor blows fuse or does not stop when switch is turned to OFF position.	Grounded near switch end of winding. Shorted or grounded armature winding.
Jerky operation, severe vibration.	High mica between commutator bars. Dirty commutator or commutator is out of round. Worn brushes and/or annealed brush springs. Open circuit or short circuit in the armature winding. Shorted or grounded armature winding.

CHAPTER 10

Commutators and Brushes

An electric motor consists of a number of parts. The shape and function of each part are determined by the type of motor and its practical application. Armatures are found in DC motors and universal motors. The wound rotor or armature is used in both DC and AC motors. If the rotor is wound and the motor operates on AC, it is most likely a universal motor and its application will be that of a small hand drill, sander, or similar device. The universal motor is also used in vacuum cleaners. Some types of heavy-duty industrial motors use a wound rotor. However, they are not included here since they have special applications and are made in sizes larger than 1 horsepower.

Fractional-horsepower motors that burn out are replaced rather than repaired. However, the price of new motors is becoming such that people are considering learning more about rebuilding rather than replacing small motors. Most of the armatures used in hand drills, saws, and sanders are machine wound, and are almost impossible to rewind. However, some maintenance can be performed on them to make sure the commutator segments are in proper shape and the brushes are making good contact with the commutator surface. Dirt and lint removal is also to be considered a necessary part of the maintenance program for small motors.

Before you throw out and replace a small motor, it is possible to check it over to learn just what is wrong with the armatures. There may be broken wire that can be soldered, or some other problem may be visible and easily corrected. Remember, the armature is connected in series with the field coils, and if there is an open circuit anywhere within the motor wiring, the circuit is not completed and the motor will not run.

The Armature

The armature is made up of a shaft, commutator, armature poles or core, laminations (called teeth), and copper wire that is wound around the pole pieces.

In Fig. 10–1 the armature is stripped and ready for rewinding. Note that in Fig. 10–2 the bottom lead of the coil of wire that has been wound around the armature poles is brought out and soldered to the commutator segment.

In Fig. 10–3 a completely wound armature is shown. The entire procedure of winding an armature will be discussed and illustrated later in this chapter.

Magnet-Wire Characteristics

Today, magnet wire is used to rewind stators and armatures. It is a quality product with a coating that can be trusted to bend (with the wire) without creating damaging cracks that can later develop into trouble spots.

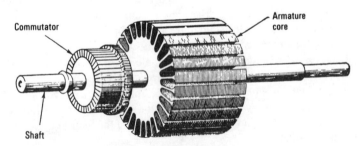

Commutator

Armature core

Shaft

Fig. 10–1. A stripped commutator ready to be rewound.

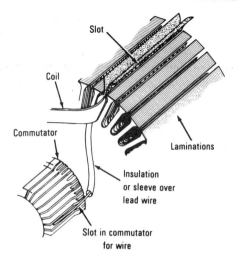

Fig. 10–2. Connection of the coil lead to the commutator.

At one time, the wire available for motor winding was coated with cotton fabric. In some cases it was double-coated cotton (dcc). This was soaked in an insulation solution after winding to produce a winding that would take the wear. Then came a clear varnish and then a black varnish with asphalt as one of its ingredients. It worked very well with round wire.

In 1938, polyvinyl formal (FORMVAR) was introduced. It is the primary type of insulation used on today's magnet wire employed in the winding of motors in fractional-horsepower sizes. There are about

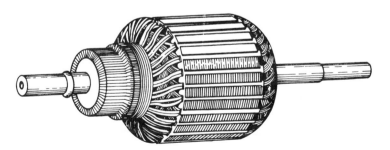

Figure 10–3. A wound armature.

300 types of coatings available for magnet wire. The Japanese use polyethylene for a coating on magnet wire used for submersible pump motors. The magnet wire is first coated with polyethylene and then topped with an overcoat of nylon.

Table 10–1 shows the turns per square inch of magnet wire per size. It also indicates the diameter of the magnet with FORMVAR insulation.

The insulation is no more than 1 to 2 mils (0.001 to 0.002 inch) in

Table 10–1. Turns per Square Inch of Magnet Wire and Diameter of the Wire with FORMVAR Insulation

Size AWG	Turns per Square Inch	Diameter of Magnet Wire
16	361	52.6
17	453	47.0
18	570	42.0
19	711	37.5
20	893	33.5
21	1110	30.0
22	1400	26.8
23	1760	23.9
24	2200	21.3
25	2740	19.1
26	3440	17.0
27	4270	15.3
28	5370	13.6
29	6760	12.2
30	8530	10.8
31	10600	9.7
32	13100	8.8
33	16500	7.8
34	20400	7.0
35	25900	6.2
36	31900	5.6
37	40800	5.0
38	50200	4.5
39	64700	4.0
40	79600	3.6

Diameter is given in *mils*, or thousandths of an inch.
Single FORMVAR insulation is shown in the diameter column.

thickness. However, in some cases it may reach 15 to 20 mils (0.015 to 0.020 inch). It is best to remember, when checking the size of wire with a wire gage, to allow one size for the thickness of the insulation. For instance, if you measure the wire and find it is No. 17 on the wire gage, take into consideration the thickness of the insulation and consider the actual size of the wire as No. 18. That is, the copper wire diameter was No. 18 before the insulation was added.

Today's insulations are capable of long-term operation at higher temperatures than those with double-coated cotton insulation. It does not mean, however, that the motor can be operated for long periods overheated. The heat buildup in the armature or the stator coils will cause the insulation to break down and damage the winding permanently, requiring a rewind of the coils or armature.

Table 10–2 shows the wire size in a number of insulation thicknesses, bare wire diameter, weight for 1000 feet, area in circular mils, and resistance per 1000 feet at 77°F (25°C). This should be sufficient for checking any type of insulation found on old and new motors of fractional horsepower size.

Fig. 10–4 shows how magnet wire is wound one turn on top of the other inside the slot of a motor's stationary pole piece. Note how the wire is insulated from touching the metal wall by a piece of insulating material.

Types of motors other than DC and universal also need wound armatures. Fig. 10–5 shows the stripped repulsion-induction motor armature. Keep in mind that the explanation of this type of motor is included here to illustrate its type and not because it is a motor which you will be encountering. This is an industrial-type motor and is made in large-horsepower sizes.

Notice the location of the squirrel-cage winding on the armature. Little current flows in the windings of this type of armature during starting moments. The heavy-gage shorting bar pointed out in Fig. 10–5 has very low inductance—the coil has very high inductance. Inductive reactance keeps the current low in the winding but very high in the shorting squirrel-cage. This means that the shorting ring has high current and dominates the effect of the total armature in the circuit. The motor will start the same as any simple repulsion motor. Once it is started, however, and normal speed is attained, the inductance of the squirrel-cage winding decreases. The current flowing in the shorting ring decreases, but not to such an extent that it is having no effect on

Table 10–2. Wire Table—Characteristics of Magnet Wire for Motor Winding

AW Gage No.	Diameter, Bare	Diameter Over Insulation								Weight (pounds per 1,000 feet)				Area in Circular Mils	Resistance in Ohms per 1,000 feet at 77°F(25°C)
		Enam. or Single FORMVAR	Double FORMVAR	Quadruple FORMVAR	Single Paper Enam.	Single Cotton Enam.	Single Silk (or Nylon) Enam.	Double Glass Bare	Single Glass Heavy FORMVAR	Bare	Double FORMVAR	Single Cotton Enam.	Double Glass Bare		
13	0.072	0.0738	0.0753	0.0779	0.077	0.0786	—	0.0795	0.0803	15.68	15.9	16.0	16.3	5178	2.043
14	0.064	0.0659	0.0673	0.0699	0.0691	0.0707	—	0.0716	0.0723	12.43	12.6	12.7	13.0	4107	2.575
15	0.057	0.0588	0.0602	0.0628	0.062	0.0637	0.0604	0.0646	0.0652	9.858	10.05	10.1	10.3	3257	3.247
16	0.051	0.0525	0.0539	0.0563	0.0557	0.0571	0.0541	0.0583	0.0589	7.818	7.96	8.06	8.24	2583	4.094
17	0.045	0.0469	0.0482	0.0506	0.0501	0.0515	0.0485	0.0528	0.0532	6.20	6.34	6.41	6.56	2048	5.163
18	0.040	0.0418	0.0432	0.0456	0.045	0.0466	0.0434	0.0478	0.0482	4.917	5.02	5.10	5.23	1624	6.51
19	0.036	0.0374	0.0386	0.0409	0.0406	0.0421	0.039	0.0434	0.0436	3.899	4.00	4.06	4.18	1288	8.21
20	0.032	0.0334	0.0346	0.0368	0.0366	0.038	0.035	0.0395	0.0396	3.092	3.17	3.24	3.34	1022	10.35
21	0.0285	0.0299	0.031	0.0331	0.0331	0.0345	0.0315	0.036	0.0360	2.452	2.51	2.57	2.67	810.1	13.05
22	0.0254	0.0267	0.0278	0.0298	0.0299	0.0313	0.0283	0.0328	0.0328	1.945	1.99	2.05	2.14	642.4	16.46
23	0.0226	0.0239	0.0249	0.0269	0.027	0.0284	0.0254	0.0301	0.0299	1.542	1.58	1.63	1.72	509.5	20.76
24	0.0201	0.0213	0.0224	0.0243	0.0245	0.0259	0.0229	0.0276	0.0274	1.223	1.26	1.31	1.39	404.0	26.17
25	0.0179	0.0191	0.0201	0.022	0.0215	0.0233	0.0207	0.0232	0.0233	0.9699	0.998	1.04	1.12	320.4	33.00
26	0.0159	0.0170	0.018	0.0198	0.0194	0.021	0.0186	0.0212	0.0213	0.7692	0.793	0.837	0.900	254.1	41.62
27	0.0142	0.0153	0.0161	0.0178	0.0177	0.0193	0.0169	0.0195	0.0194	0.610	0.630	0.662	0.727	201.5	52.48
28	0.0126	0.0136	0.0145	0.016	0.016	0.0178	0.0152	0.0179	0.0177	0.4837	0.501	0.532	0.588	159.8	66.17
29	0.0113	0.0122	0.013	0.0145	0.0146	0.0164	0.0138	0.0166	0.0162	0.3836	0.396	0.427	0.477	126.7	83.44
30	0.0100	0.0109	0.0116	0.0131	0.0133	0.0151	0.0125	0.0153	0.0149	0.3042	0.316	0.344	0.387	100.5	105.2
31	0.0089	0.0098	0.0105	—	—	0.0139	0.0113	0.0142	0.0137	0.2413	0.251	0.278	0.314	79.7	132.7
32	0.0080	0.0088	0.0094	—	—	0.013	0.0104	0.0133	0.0126	0.1913	0.198	0.224	0.255	63.71	167.3
33	0.0071	0.0079	0.0085	—	—	0.0119	0.0094	0.0124	0.0117	0.1517	0.158	0.182	0.208	50.13	211.0
34	0.0063	0.0071	0.0075	—	—	0.0111	0.0086	0.0116	0.0108	0.1203	0.126	0.148	0.169	39.75	266.0
35	0.0056	0.0063	0.0067	—	—	0.0103	0.0078	0.0109	0.0100	0.0954	0.0996	0.120	0.138	31.52	335.5
36	0.0050	0.0057	0.0060	—	—	0.0093	0.0072	0.0103	0.0093	0.0757	0.0791	0.100	0.113	25.00	423.0
37	0.0045	0.0051	0.0055	—	—	0.0087	0.0066	0.0098	0.0087	0.060	0.0628	0.080	—	19.83	533.4
38	0.0040	0.0045	0.0049	—	—	0.0082	0.0061	0.0093	0.0081	0.0476	0.0498	0.068	—	15.72	672.6
39	0.0035	0.0040	0.0043	—	—	—	0.0055	0.0088	0.0075	0.0377	0.0397	0.060	—	12.47	848.1

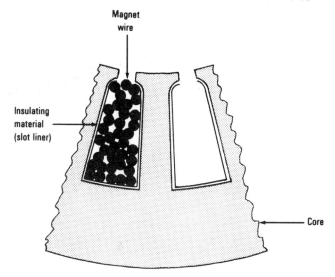

Fig. 10–4. Placing the magnet wire coil into a slot in the stator of a motor.

the operation of the motor. It still dominates over the wound coil. That accounts for the characteristics of this type of motor. No shortcircuiting device is needed as is the case in the repulsion-start motor. The characteristics of this type of motor are high starting torque and constant speed. It is not the type of motor found in appliances around the home,

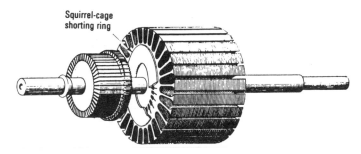

Fig. 10–5. Squirrel-cage shorting ring on a rotor designed for a repulsion-induction motor.

but is primarily a special-application industrial motor. It does require maintenance since it has brushes and a wound rotor.

Commutator Maintenance

The most important factor, and the one on which the success or failure of a DC motor and commutator-type AC motor depends, is commutation. Satisfactory commutation means operating under reasonable conditions without excessive sparking, burning of commutator bars or brushes, or other conditions requiring excessive maintenance.

Assuming that the design of a machine is such that good commutation is to be expected, one of the best means of securing satisfactory operation operation is through maintaining the surface of the commutator in good operating condition. Generally, this means that the commutator surface should be smooth, concentric, and properly undercut.

Resurfacing

There are three methods used in truing commutators. They are sandpapering, hand stoning, and grinding or turning. Hand stoning is used, or was used, more extensively in large motors. Sandpapering is used on small commutators. Grinding or turning can be used on small and medium-size commutators that can be mounted between the tail stock and chuck on a lathe. The method used depends on the degree of damage and how you are equipped to handle the commutator.

Sandpapering—This is a satisfactory method of removing deposits from a commutator surface. It can be used to correct roughness or reduce high mica provided the accurate contour of the surface has not been disturbed. However, sandpaper cannot be depended on to remove flat spots even of small size.

One of the principal objections to the use of sandpaper for cleaning commutators is that it very rarely leaves the bars properly round. Oil should not be used when sanding a slotted commutator. It will cause the copper dust and sand to collect in the slots, while if no oil is used, centrifugal force will throw the dust from the slots.

Fig. 10–6 shows how a piece of sandpaper can be attached to a curved block to aid in the proper sanding of a commutator. *Caution: Emery paper or emery cloth should never be used on a commutator.*

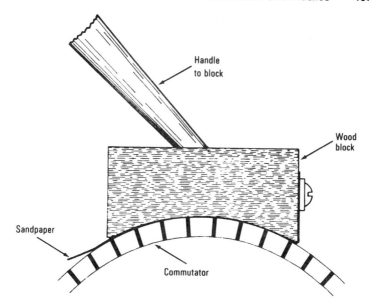

Fig. 10–6. Making a sanding block for smoothing commutator surface. Used on very large DC motors.

Emery is an electrical conductor and particles are likely to become embedded between the segments and cause short circuits.

Hand Stoning—The tendency of sandpaper to broaden rather than remove flat spots and to distort the contour of the commutator segments results from the flexibility of the paper. The use of a commutator stone is, therefore, recommended in preference to sandpaper. This is, of course, for larger DC motors and generators designed for industrial use. In small fractional-horsepower motors, it is unnecessary to stone the commutator. Various sizes of commutator hand stones are made for use on larger commutators (Fig. 10–7). The stone is pressed firmly against the commutator and moved slowly from side to side.

Grinding or Turning—While sandpapering offers a satisfactory method of resurfacing a commutator which is dirty or on which the mica is just beginning to build up, hand stoning will do a better job of removing high mica and even flat spots of considerable size. The most

Fig. 10–7. Hand-held stones for turning down a commutator. Used on very large DC motors.

satisfactory method of resurfacing a commutator that is badly out of round is by grinding or turning with a tool having a rigid support. All cases of eccentricity come under this head, for no hand method of finishing can entirely eliminate eccentricity in a commutator.

When truing the commutator in a lathe, you can support the armature on the lathe centers or on sleeve bearings. The former is simpler but the latter method has some advantage in point of accuracy.

Before starting to turn or grind a commutator, the windings of the armature (adjacent to the commutator) should be wrapped in cloth or plastic to protect them from copper chips and dust. In turning a commutator with a steel tool, use what is known as a diamond point tool or one with a very sharp point. The point of the tool should be rounded sufficiently that the cuts will overlap and not leave a rough thread on the commutator surface. Use a light cut, taking several cuts if necessary to remove bad flat spots or a considerable degree of eccentricity.

After turning the armature's commutator, it is a good idea to use a very fine sandpaper and hold it on the surface while the commutator is turning. Then clean the dust from the slots between the commutator segments.

Undercutting Commutator Mica

The object of undercutting of commutator mica is to remove the mica between the copper segments so that the segments will wear evenly. The removal of the mica is necessary.

Fig. 10–8A shows how a hacksaw blade has been used to undercut the mica insulation between the commutator segments. The blade may have to be ground to fit the width of the slot.

Fig. 10–9 shows two correct ways and two incorrect ways to undercut the mica. Keep in mind the main reason for doing this is to make the motor operate smoothly. If the mica is sticking up above the surface of the copper, the brushes cannot make contact with the commutator segments.

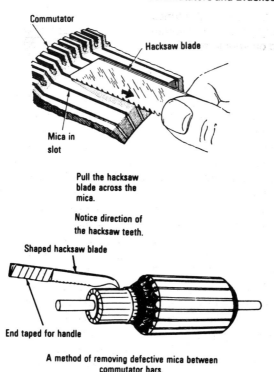

Commutator

Hacksaw blade

Mica in slot

Pull the hacksaw blade across the mica.

Notice direction of the hacksaw teeth.

Shaped hacksaw blade

End taped for handle

A method of removing defective mica between commutator bars.

Fig. 10–8. (A) Using a hacksaw blade to cut the high mica on a commutator.

Be sure to clean the commutator before replacing it in the motor, and make sure that under no conditions are grease and oil allowed to become part of the commutator surface. Small electrical tools have pitted commutator segments and may need a slight sanding to bring them back into normal operation. The quick way to check is to look at the motor while it is operating and see if there are many sparks where the brushes touch the commutator. There is supposed to be some sparking or arcing; the trick is to learn what is the correct amount for normal operation. If the commutator is pitted and black, it probably needs to be resurfaced or sanded. It is a good idea to check every time you change brushes in the motor. Most smaller motor manufacturers suggest that you inspect the commutator for wear each time you replace

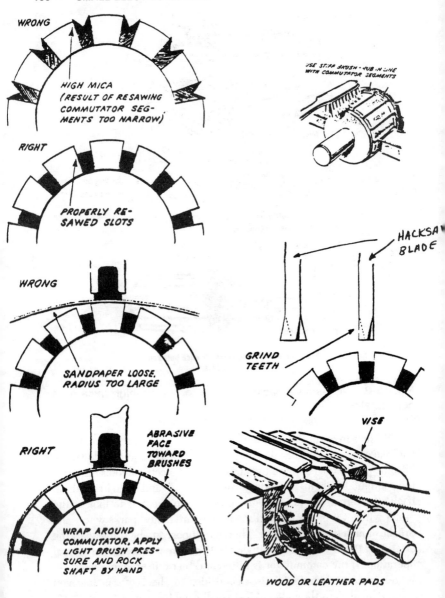

WRONG

HIGH MICA
(RESULT OF RESAWING
COMMUTATOR SEG-
MENTS TOO NARROW)

RIGHT

PROPERLY RE-
SAWED SLOTS

WRONG

SANDPAPER LOOSE,
RADIUS TOO LARGE

RIGHT

ABRASIVE
FACE
TOWARD
BRUSHES

WRAP AROUND
COMMUTATOR, APPLY
LIGHT BRUSH PRES-
SURE AND ROCK
SHAFT BY HAND

USE STIFF BRUSH - RUB IN LINE
WITH COMMUTATOR SEGMENTS

HACKSAW
BLADE

GRIND
TEETH

VISE

WOOD OR LEATHER PADS

Fig. 10–8. (B) Working on commutators.

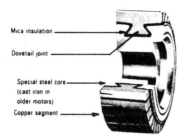

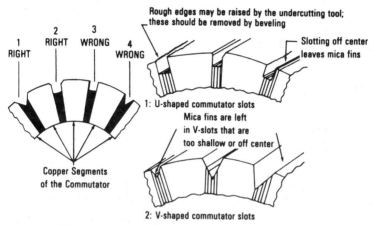

Fig. 10–9. Right and wrong ways to cut high mica.

the brushes. If the commutator is worn down more than $\frac{1}{32}$ inch (0.80 mm) in diameter, turning and undercutting is recommended. Usually three sets of brushes can be used for one commutator turning.

Brushes

Brushes are necessary to complete the path for current flow through the armature, which puts some very heavy demands on them. They must be smooth enough in their contact with the copper conductor, yet not have oil or lubricant on them, to glide over without causing undue damage; they have to make good electrical contact; and they have to be able to handle the current needed in the specific situation where they are used. Current density is the current-carrying capacity of brushes expressed in amperes per square inch of useful brush area in normal contact with the commutator surface. The current density varies with the application. The highest density is about 150 amperes per square inch for certain super-baked grades.

Brushholders

Brushes must be held securely to the surface of the commutator, which usually requires spring action. The spring tension is supplied by a screw-in type or cap or by a coiled spring mounted on the outside of the motor (Fig. 10–10).

Brush Types

There are four popular types of brush materials. The *carbon-graphite brush* is made of hard carbon graphite and is particularly well suited for use with motors operating with flush mica commutators,

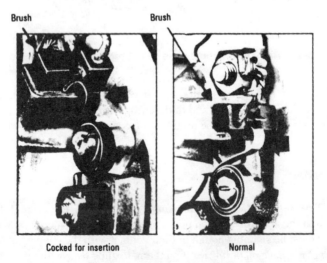

Brush   Brush

Cocked for insertion   Normal

Fig. 10–10. Brush cocking procedure for a permanent-magnet DC motor. *(Courtesy Doerr)*

where appreciable polishing action is required. The density is 35 to 45 amperes per square inch, which means it is used only on low-current fractional-horsepower motors.

The *electro-graphite brush* is made by subjecting carbon to intense heat. That means its conversion to graphite is a physical, not a chemical, change. It is useful in higher-speed motors. It is less abrasive than carbon graphite, and is tougher and has greater current density—75 amperes per square inch.

Natural graphite is a mined product, just like coal. *Artificial graphite* is made in an electric furnace. It is the artificial graphite that is most often used for brushes in motors. This type of brush has a more polishing effect than electro-graphite grades. This type of brush gives good riding qualities and can be used on extremely high-speed devices, such as series motors used in routers and drills. The current density of this type of brush is in the range of 50 to 65 amperes per square inch. Artificial graphite brushes are found in a large variety of small appliances where sparking may be a problem. They are quieter than other types of brushes.

Metal-graphite brushes are made of copper and graphite mixed according to the demand for certain characteristics. This type of brush can carry extremely high currents. They have a current density of 150 amperes per square inch. That is, of course, if the copper content of the brush is 50 percent.

Selecting the Right Brush

There are a number of factors which work together to make a special brush necessary for a particular job. For example, series motors often gain higher efficiency with a low ampere-turns ratio, but with the normally used grade of brush, sparking becomes more noticeable and the commutator is blackened or burned. A proper substitute brush choice for that use would be a hard grade of brush with a slight cleaning action.

Frequent starting and stopping of a motor causes some heavy loads for brushes. When you select a brush, there is no easy method to use to make sure you are getting the right one. Recommendations from the brush and motor manufacturer will help narrow the choice of brush selection. Starting and stopping frequently, reversing, overload capacity, vibration, and brush noise are all factors that enter into the selection of the right brush for a given motor. Manufacturers of motors have most likely been busy testing their motors under various load conditions and will recommend the proper type of brush for their machines. Even humidity and temperature having a bearing on brush life and operation. Refer to the data supplied by the manufacturer of the motor for the best possible brush to use in a motor.

Fig. 10–11 shows the location of the brushes in a portable saw used in general carpentry work. In Fig. 10–12 you can see the location of brushes in a popular hand drill. Look through these holes to observe the degree of arcing. This arcing will indicate whether or not the commutator needs to be refinished.

Brush Seating

For lowest possible friction between the commutator and brushes, and for optimum commutation of armature current, an adherent and

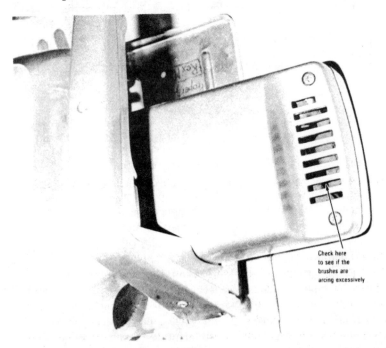

Check here
to see if the
brushes are
arcing excessively

Fig. 10–11.   Brushes are located at this end of the saw.

CHECK
BRUSHES
HERE

Fig. 10–12. Check for arcing on the electric drill near the brushes.

uniform film must be developed on the commutator surface (Fig. 10–13). To accomplish this, brushes must first be "run in" or seated. During this running period, brush material is transferred to the commutator by the direct wiping motion of the brushes and by electrolytic action. As this occurs, the initial contact points are gradually worn away and the brushes begin to conform to the curvature of the commutator surface. Softer-grade brush materials can be seated faster since this initial process is mostly mechanical.

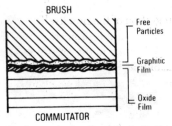

Fig. 10–13. Brush to commutator contact. *(Courtesy Bodine)*

An adherent and uniform film must be developed on the commutator surface.

Electrical Brush Wear

After the brushes have been fully seated, electrical wear surpasses mechanical wear as the most important factor in brush operation. The passage of current through the brush and intervening film causes a high electrical wear rate. Very high temperatures at the minute conduction points cause copper melting and even carbon vaporization when the arc voltage is high enough. Electrical wear is most evident at the negative, or cathode, brush. Here, the wear rate can be two or more times greater than at the anode (+) brush (Fig. 10–14).

Overall Wear Rate

Mechanical and electrical wear contribute to the overall wear rate of brushes. This rate may increase or decrease by a number of additional factors such as humidity. With a dew point below approximately −10°C, or an absolute humidity of less than 0.13 grams of water vapor per cubic foot, the commutator film will dry out. This can occur independently of operating altitude. Loss of the normal water vapor com-

Fig. 10–14. Wear at the cathode terminal or brush is always greater than at the anode (+) brush. *(Courtesy Bodine)*

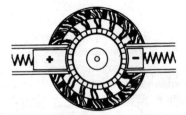

Wear rate at the cathode (−) brush can be two or more times that at the anode (+).

ponent of the film gradually leads to the breakdown and eventual disintegration of the film itself.

Without this protective film, friction can increase from two to five times, leading to the most rapid brush wear rate, known as "dusting." Dusting at high altitudes or in dry ambient air can be corrected by rehumidifying the atmosphere, or by using brushes with special additives.

Vibration from a connected load or other external source can also cause rapid brush wear. Excessive vibration will affect the way the brush rides on the commutator surface, causing it to conduct intermittently or *spark*. Intermittent contact can greatly accelerate electrical wear.

Fit, or the combination of clearance in the brushholder and brush spring pressure, is also critical to brush performance. If the brush is too loose in the brushholder or the brush spring pressure is too light, premature electrical wear can result. However, when brush pressure is too strong, heavy mechanical wear can be expected.

Commutator Insulation

If not properly undercut, the mica insulation between commutator bar segments can break off in small abrasive particles (Fig. 10–15). These particles will rapidly wear down most brush grades now in use.

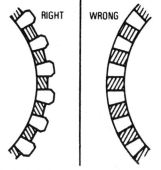

RIGHT   WRONG

Fig. 10–15. Proper and improperly cut mica on a commutator can affect brush life. (*Courtesy Bodine*)

Mica insulation between commutator bars must be undercut (Mica—cross hatched in the above diagram—is shown greatly exaggerated in width for purposes of illustration)

(1) PITCH BAR-MARKING produces low or burned spots on the commutator surface that equals half or all the number of poles on the motor.

(2) STREAKING on the commutator surface denotes the beginning of serious metal transfer to the carbon brush.

(3) HEAVY SLOT BAR-MARKING involves etching of the trailing edge of commutator bar in relation to the numbered conductors per slot.

(4) THREADING of commutator with fine lines is a result of excessive metal transfer leading to resurfacing and excessive brush wear.

(5) COPPER DRAG is an abnormal amount of excessive commutator material at the trailing edge of bar. Even though rare, flashover may occur if not corrected.

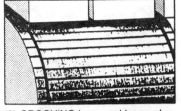

(6) GROOVING is caused by an abrasive material in the brush or atmosphere.

Fig. 10–16. Commutator wear and markings. *(Courtesy of Reliance Electric)* **(A) Types of problems.**

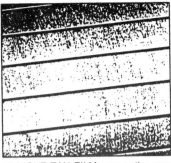

(1) LIGHT TAN FILM over entire commutator surface is a normal condition.

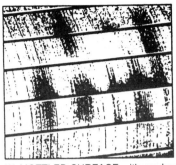

(2) MOTTLED SURFACE with random film patterns is satisfactory.

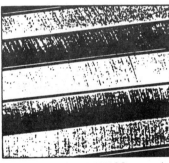

(3) SLOT BAR MARIKINGS appearing on bars in a definite pattern depicts normal wear.

(4) HEAVY FILM with uniform appearance over entire commutator surface is acceptable.

Fig. 10–16. (B) Examples of normal wear.

If they become embedded in the brush faces, the mica particles can also scrape across the commutator. This makes grooves in its surface.

Oil on the brushes or commutator, sometimes picked up in the film from the air, acts as an insulator. It can also act as an adhesive to hold mica particles of dirt and cause abrasive wear to the brushes and commutator. For this reason, special care must be taken when lubricating sleeve-bearing motors. Overoiling will often adversely affect brush life. Fig. 10–16 shows examples of commutator wear from various sources.

Airborne Contaminants

Airborne contaminants, such as the decomposition products of silicones, can also cause poor commutation and excessive brush wear rates. When adhered to the commutator, they act like an abrasive insulating film. Other gaseous substances such as methyl alcohol can reduce the copper oxide in the commutator film and increase brush friction.

Contaminants such as turpentine, paint fumes, and acetone can increase brush friction and accelerate brush wear. Other harmful vapors come from chlorinated hydrocarbon solvents and acid-forming gases such as sulfur dioxide and hydrogen chloride. Even tobacco smoke can raise friction enough to double the brush wear rate.

There is no magic formula for selecting a suitable brush to combat these effects. Motor and brush manufacturers work to select optimum brushes for a given application and environment.

Brush Maintenance

To get the best performance and longest life from motor brushes, the user must develop a regular maintenance procedure. This should include:

1. Periodic inspection of brushes, and replacement when they have worn down to ¼ inch (7 mm).
2. Inspect brush springs and replace when damaged or collapsed.
3. Periodic cleaning and removal of built-up dust inside the motor. The decomposing or worn brush will deposit its dust.
4. The motor should always be disconnected from the power source before inspecting or replacing the brushes.

CHAPTER 11

Shafts and Bearings

All motors have shafts and bearings. The selection can be made for a long shaft or a short one or any size in between. The shaft length, finish, diameter, and design are all subject to the demands of the buyer. Some general types are available for fans, pulleys, and direct drives which are lubricated by bearings. The bearings may be either of the sleeve or ball design; each has its advantages and disadvantages. Lubrication becomes a very important part of maintaining a motor. The type of lubricant and the frequency of application determines, in most instances, where the motor will be useful. Therefore, it is very important for the buyer to select the proper bearings and shaft configuration for the job to be done.

Motor Shafts

The unit heater motor shown in Fig. 11–1 has a shaft with a flat spot. The flat can be used to make sure the setscrew in the pulley will not slip and eat into the shaft. Some pulleys have a flat spot in their design to fit directly over this type of shaft.

Some motors have one end of the shaft that extends past the body of the motor. Others have an extended shaft on both ends (Fig. 11–2).

Fig. 11–1. Unit heater motor, enclosed, capacitor-start, with sleeve bearings and resilient mounting. *(Courtesy Westinghouse)*

Fig. 11–2. A fan coil motor with rubber end pieces to absorb vibration when mounted. Hexagonal holes fit over the end of the motor. The motor is thermally protected. *(Courtesy Westinghouse)*

This will allow for attaching two separate devices to the motor. It also increases the load on the bearings. Fig. 11–3 shows how the shaft extends through the rotor. It is usually ground to a precision fit and is either press-fit onto the rotor or welded in place. Note how the shaft has been machined to fit the bearing surfaces and the pulley it will eventually drive.

Gearmotors have short shafts since the end of the shaft has been grooved to match the gears in the gearbox. See Fig. 11–4 for the short shaft that drives a gearbox for the reduction of the motor speed. The gearbox is bolted onto the motor and the gears mesh to provide a greater torque at a slower speed for a special application.

A typical single reduction right-angle gearmotor is shown in Fig. 11–5. Here the shaft is a little longer and a gear is mounted on its end

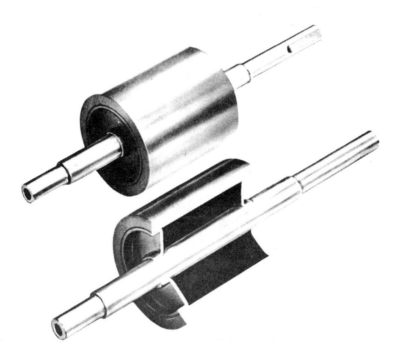

Fig. 11–3. The shaft is machined to fit the bearings and to be press-fit through the rotor. *(Courtesy Bodine)*

Fig. 11–4. Gearbox with a straight-through shaft. *(Courtesy Bodine)*

to mesh with the gearbox right-angle shaft. Note how the bearings are placed and lubricated.

Rotor and Shaft

To present a typical specialty-motor design characteristic, let's take a look at Fig. 11–6. This is the rotor and shaft from a Westing-

Fig. 11–5. A gearbox with a 90° drive shaft. *(Courtesy Bodine)*

Fig. 11–6. A shaft with a flat side for setscrews. *(Courtesy Bodine)*

house Frame 42 specialty motor. Precision die-casting equipment forces molten primary aluminum through the rotor core to form the rotor conductor bar cage. The preheated rotor is accurately located on the finished shaft and allowed to cool to a tight shrink-fit. Precision rotor machining and uniform aluminum die casting ensure excellent rotor balance and concentricity.

A number of rotor skew options are available to reduce noise and vibration. Shafts are precision-finished to exacting tolerances to prevent or reduce bearing wear. Phosphate is used to treat each carbonsteel shaft for protection against rust and corrosion. A clean corrosionfree shaft ensures easy removal of shaft-mounted devices.

Shaft Adapters

The motor you have may not have the correct size shaft or the right slot mechanism. This can be remedied by obtaining a transition shaft adapter (Fig. 11–7). Two sizes of shaft adapters with keys are available for use with motors where the shaft diameter of the new motor is smaller than the shaft of the motor being replaced. One adapter will increase the shaft diameter from ½ inch to ⅝. The other adapter will change the shaft diameter from ⅝ inch to ¾ inch.

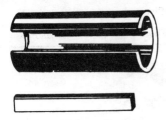

Fig. 11–7. Transition shaft adapter. *(Courtesy General Electric)*

Selection of Bearings

It is extremely important to select a motor or gearmotor with the correct bearings. Severe conditions of operation dictate the type of bearing to be used. Since metal-to-metal contact during rotation causes friction and heating, care in the selection of a drive unit with the appropriate bearings for the intended application is an essential factor in the life and effectiveness of any driven machine. Among the many considerations which affect the selection of bearings are speed requirements, temperature limits, lubrication, load capacity, noise and vibration, tolerance, space and weight limitations, end thrust, corrosion resistance, infiltration of dirt or dust, and, of course, cost.

There are two principal types of bearings to fit the needs of the various motors under varying conditions. Fractional-horsepower motors use sleeve (journal) and ball bearings.

In gearheads for fractional-horsepower motors you will find sleeve, ball, tapered-roller, needle-thrust, and draw-cup full-complement needle bearings.

Characteristics of ball and journal (sleeve) bearings are given in Table 11–1. This data and the application data will aid in the selection of the proper motor.

Sleeve Bearings

Sleeve, or journal, bearings are the simplest in construction. They are the most widely used when low initial cost is a factor. They are quiet in operation and have a good radial load capacity. They can be used over a fairly wide temperature range. Sleeve bearings also have virtually unlimited storage life if the motor is to remain unused for extended periods of time. They show good resistance to humidity, to mild

Table 11–1.   Comparison of Ball- and Sleeve-Bearing Characteristics

Characteristics	Sleeve Bearing	Ball Bearing
Load		
Unidirectional	Good	Excellent
Cyclic	Good	Excellent
Starting and Stopping	Poor	Excellent
Unbalanced	Good	Excellent
Shock	Fair	Excellent
Thrust	Fair	Excellent
Overhung	Fair	Excellent
Speed Limited By	Turbulence of oil. Usual limit (5000 rpm max.)	20,000 rpm max.
Misalignment Tolerance	Poor (unless of the self-alignment type)	Fair
Starting Friction	High	Low
Space Requirement		
Radial	Small	Large
Axial	Large	Small
Damping of Vibration	Good	Poor
Type of Lubrication	Oil	Oil or Grease
Lubricant, Amount Required	Large	Small
Noise	Quiet	Depends upon quality of bearing and resonance of mounting
Low Temperature Starting	Poor	Good
High Temperature Operation	Limited by lubricant	Limited by lubricant
Maintenance	Relubrication required periodically	Relubrication required only occasionally. Grease-lubricated bearings often last for the life of the application without attention

(Courtesy Bodine)

dirt infiltration, and to corrosion if made of bronze. Under light loads, the static friction of sleeve bearings is very nearly as low as grease-packed ball bearings (Fig. 11–8).

The main disadvantage of the sleeve bearing is the need for relubrication. They are longer than ball bearings and may be somewhat at a disadvantage if space or size is a factor. An oil reservoir, felt, or similar oil-retaining material must also be incorporated into the end shield and the lubricating oil must be replenished periodically (Fig. 11–9). *Caution: Sleeve bearings cannot be allowed to run dry!*

Graphitized Self-Lubricating Sleeve Bearings—Graphitized self-lubricating sleeve bearings are made of solid bronze. They have graphite as an inner recess filler. The recesses are often in the shape of two figure-eights. They may also employ graphite-filled holes to conduct oil between the reservoir and the inner bearing surface. The bronze bearing or body of such bearings provides strength and resistance to shock or vibration, while the presence of graphite helps form a lubricating film on the bearing surface. It also prevents metal-to-metal contact when the motor is stopped. The graphite will also act as an emergency lubricant if the oil level is allowed to run low. Graphited bearings will also usually withstand higher operating temperatures than ordinary sleeve bearings.

Porous Bronze Sleeve Bearings—Another type of self-lubricating bearing (sleeve type) is made from porous bronze. The porous bronze sleeve bearing is oil-impregnated and can be used with a felt washer around its periphery to hold additional oil in suspension, making frequent relubrication unnecessary (Fig. 11–10).

Porous bronze bearings are more compact and offer more freedom from attention than solid bronze bearings. They are often constructed

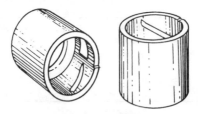

Fig. 11–8. Large, babbitt-lined sleeve bearings are precision-machined to extremely close tolerances for accurate alignment. *(Courtesy General Electric)*

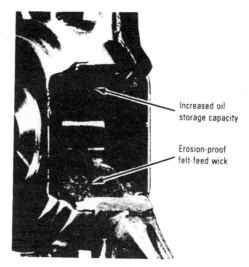

Increased oil
storage capacity

Erosion-proof
felt-feed wick

Fig. 11–9. Oil reservoir for a sleeve-bearing motor. *(Courtesy General Electric)*

to be self-aligning, to reduce friction and to be shaft-binding. The porous bearing is generally more economical than the graphited or solid bronze types.

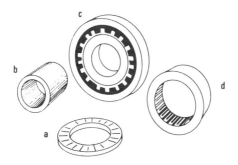

Fig. 11–10. Comparison of bearing types: (A) Needle-thrust bearing; (B) sleeve bearing; (C) ball bearing; (D) full-complement drawn-cup needle bearing. *(Courtesy Bodine)*

Sleeve-Bearing Lubrication

Lubricants for sleeve bearings must provide an oil film that completely separates the bearing surface from the rotating shaft member and ideally eliminates metal-to-metal contact. Oil, because of its adhesion properties and its viscosity (or resistance to flow), is dragged along by the rotating shaft of the motor and forms a wedge-shaped film between the shaft and the bearing (Fig. 11–11). The oil film forms automatically when the shaft begins to turn and is maintained by the motion. The rotational motion sets up a pressure in the oil film wedge, which, in turn, supports the load. That means the oil selected to carry this load is very important.

High temperatures and high motor operating temperatures will have a destructive effect on sleeve bearings lubricated with standard-temperature-range oils. Special oils are available for motor applications at high temperature, and also for applications at lower-than-normal temperatures. Motor performance and bearing life are directly related to the lubrication.

Sleeve bearings are not as sensitive to abrasive or foreign matter as are ball bearings. They are able to absorb some of the small particles in the soft undersurface of the bearings. However, good maintenance practice dictates that you keep the lubrication oil as clean as possible.

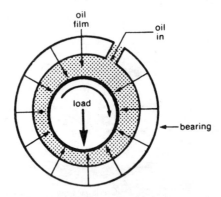

Fig. 11–11. The oil film in a hydrodynamic bearing. A wedge-shaped gap is formed between the shaft and bearing. Higher pressures are developed where the gap narrows, which lifts the shaft and its load. *(Courtesy Bodine)*

A conservative lubrication and maintenance program should call for periodic inspection of the oil level and cleaning and refilling with new oil every six months.

Note: Sleeve-bearing motors may tend to lose their oil film when stored for extended periods (one year or more).

Ball Bearings

Virtually all sizes and types of electric motors use ball bearings. They have low friction loss when lubricated with oil, are suited for high-speed operation, and can be used in relatively wide ranges of temperature.

Ball bearings can accommodate thrust loads and permit end play to be conveniently minimized. Compared to sleeve bearings, ball bearings require significantly less maintenance, especially if grease-packed (Fig. 11–12).

Ball bearings are slightly more expensive than sleeve bearings. They are also noisier. This is due to the nature of their rolling action. The latest developments have noticeably reduced the noise levels in ball-bearing motors (Fig. 11–13).

Fig. 11–12. A double-shielded ball bearing. *(Courtesy SKF Industries)*

Fig. 11–13. Cutaway view showing shielded ball bearing installed in a general-purpose cage motor. *(Courtesy Allis-Chalmers)*

This type of bearing is more susceptible to rust since it is made of steel. If it is kept in storage for some time, it may show a tightening of the shaft due to the lubricant hardening. This can also be caused by low temperatures. Giving the motor some warm-up time will, in some instances, rejuvenate the ball-bearing grease to a suitable condition. Recently, grease with long life has been developed.

Fig. 11–14 shows how the ball bearing is mounted in the motor. Compare this with Fig. 11–15, where the sleeve bearing is used. Note the differences in the lubrication methods and design of the bearings.

Ball-Bearing Lubrication

The main purpose of the ball-bearing lubricant is to keep out dirt or moisture. It also helps to dissipate the heat that builds up in the bearing, but it does not provide an oil film to reduce bearing friction.

Overlubrication of ball-bearing motors is more of a problem than is insufficient lubrication or overloading the motor. Both sleeve- and ball-bearing motors can suffer from overlubrication. Too much lubri-

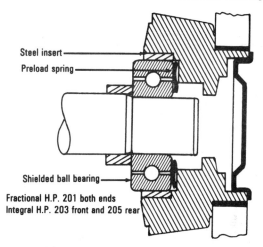

Steel insert

Preload spring

Shielded ball bearing

Fractional H.P. 201 both ends
Integral H.P. 203 front and 205 rear

Fig. 11–14. Ball bearing installed in a motor with a preload spring.
(Courtesy Westinghouse)

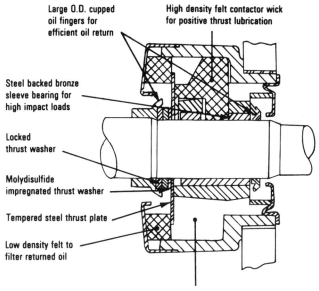

Large O.D. cupped
oil fingers for
efficient oil return

High density felt contactor wick
for positive thrust lubrication

Steel backed bronze
sleeve bearing for
high impact loads

Locked
thrust washer

Molydisulfide
impregnated thrust washer

Tempered steel thrust plate

Low density felt to
filter returned oil

Lubricating material (having high oil release rate)

Fig. 11–15. A sleeve-bearing installation in a motor. *(Courtesy Westinghouse)*

cant will eventually find its way into the stator windings and break down the windings' insulation.

Either oil or grease may be used in lubricating ball bearings. Check the manufacturer's specifications. The life of the bearings will depend on the operating conditions. Generally, relubrication of the bearings during the lifetime of the motor is not necessary.

Ball bearings can be mounted in a motor in almost any position. If the bearing has seals on both sides, it is not designed to be relubricated. Instead, if the bearing is worn or needs attention, it should be replaced.

Oil is the most efficient lubricant. Oiling does, however, require some rather elaborate oiling devices. Seals and closures are used to prevent leakage of the oil. Lubricant levels must be maintained and only good-quality oils should be used.

Sleeve bearings are less sensitive to a limited amount of abrasive or foreign materials than are ball bearings. Oil and bearings should be kept clean. In larger, industrial-type motors, the oil should be changed at regular intervals. The split-phase motor usually requires oiling the wick or placing a few drops into the cup once every six months (Fig. 11–16). *Remember: Never overlubricate the bearings because the oil will seep into the insulation and settle on the windings of the motor.*

Bearing Failure

The most common type of motor problem is bearing failure. This can lead to shaft failure as well if it is not caught in time. Sleeve-bearing replacement consists of pressing out the old bearing and pressing in a new one. This can be done with a drill press fitted with a dowel rod the

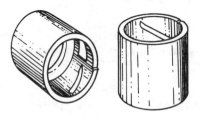

Fig. 11–16. Sleeve bearing showing oil grooves.

same diameter as the sleeve bearing. The dowel rod may have to be turned slightly to fit into the drill press chuck. However, a downward pressure from the drill press arm will cause the bearing the slip out of the bell housing of the motor. If this doesn't work, use the bearing mandrels made for the removal of bearings. A slight tap of the ball-peen hammer on the mandrel will remove the bearing.

The new (or replacement) bearing must be an exact factory replacement, or it must be installed and line-reamed for an exact fit. If not, repeated failure is certain.

Ball-bearing replacement is easier than sleeve-bearing replacement. A bearing puller and inserter is almost a necessity. Take a look at Chapter 1, on tools, to see what tools are needed. See especially Figs. 1–20 and 1–21.

Shaft Repair

Shafts may be damaged if the bearings are worn too much . If the shaft is damaged for this or any other reason, it is possible to regrind it so the bearings will fit properly. This usually takes some special skills with metalworking tools and machines. It is not advisable to undertake the task if you are inexperienced. In some instances, however, the shaft can be damaged by a pulley not fitting properly. This may require removal of some parts by a hand file or light grinding. Another pulley may be required to fit the refinished surface of the shaft, depending on the amount of metal removal that is involved. Not all shafts are repairable or replaceable. When some go bad, the whole armature or rotor must be replaced.

Bearings can be worn by an unbalanced armature or rotor. Fig. 11–17 shows how the armature of a hand drill is balanced by the drilling of holes to remove material in the laminations. Three-phase motors use the addition of small weights in the fan blades to help balance the rotating member of the motor (Fig. 9–16). Modern motor manufacturing techniques use computers to aid in balancing the rotor and the armature. The whole job is usually done by removing metal from the laminations as shown in Fig. 11–17. This is the easier and faster method for production purposes. The balanced rotor or armature adds considerably to the quietness of the motor. Armatures of series (or universal) motors can rotate at speeds up to 20,000 rpm. This is sufficient

Fig. 11–17. Holes drilled in the armature laminations aid in the balance of the rotating member.

to cause all kinds of trouble with vibration noise if the armature is not balanced.

Motor Problems Caused by Bearing Wear or Damage

A number of problems can be observed as being caused by bearing wear. Bearings can be worn excessively if the belt to the pulley on the shaft is too tight. This places undue pressure on the bearing surface. A severe blow to the shaft or the dropping of something on a shaft or pulley on a shaft can cause damage to the bearings. Once the bearings are damaged, it is only a matter of a few revolutions of the rotor or armature until the rotating member of the motor starts to strike parts of the stationary members. This can cause a number of symptoms and problems.

Worn Bearings Can Cause:
- all motors to run too hot after extended operation.
- all motors to have excessive mechanical noise.

Unbalanced Rotor or Armature Vibration Can Cause:

- all motors to have excessive mechanical noise.

Interference Between Stationary and Rotating Member Can Cause:

- all motors not to start.
- all motors to start, but heat up rapidly.
- all motors to have excessive mechanical noise.
- all motors to have a reduction in power and to get too hot.

CHAPTER 12

Switches and Relays

One of the most troublesome things that happens to a motor is overheating and burn-out of the insulation materials. Although motors have been designed to operate under various temperature conditions, it is always possible for a motor to overheat when a number of things take place:

1. Low line voltage increases the current drawn from the line or power source. Since an increase in current causes an increase in temperature, it is important to make sure that the motor windings are properly protected from such an increase in temperature.
2. Jamming the motor and causing it to stop when the power is on will also cause temperature increases. This takes place almost immediately and a good sensing device is needed to protect the windings.
3. Large increases in voltage can cause excess current to flow through the windings. The increased current will cause increased heat to be generated.

Overload Protection

The effect of an overload is a rise in temperature in the motor windings. The larger the overload, the more quickly the temperature will increase to a point damaging to the insulation and lubrication of the motor. An inverse relationship, therefore, exists between current and time. The higher the current, the shorter the time before motor damage, or burn-out, can occur.

All overloads shorten motor life by causing deterioration of the motor insulation. Relatively small overloads of short duration cause little damage, but if sustained, they could be just as harmful as overloads of greater magnitude. The relationship between the magnitude (percent of full load) and duration (time in minutes) of an overload is shown in the graph of Fig. 12–1.

The ideal overload protection for a motor is an element with current-sensing properties very similar to the heating curve of the motor, which could act to open the motor circuit when *full-load current* (FLC) is exceeded. The operation of the protective device should be

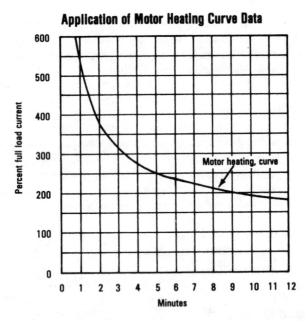

Fig. 12–1. The relationship between the magnitude and duration of an overload.

such that the motor is allowed to carry harmless overloads but is quickly removed from the line when an overload has persisted too long.

Fuses are not designed to provide overload protection. Their basic function is to protect against short circuits. A fuse chosen on the basis of motor FLC would blow every time the motor started. On the other hand, if a fuse were chosen large enough to pass the starting current or inrush current, it would not protect the motor against small, harmful overloads which might occur later (Fig. 12–2).

Dual-element or time-delay fuses can provide motor overload protection. But they suffer the disadvantage of being nonrenewable and must be replaced (Fig. 12–3).

Overload relays consist of a current-sensing unit connected in-line to the motor, plus a mechanism actuated by the sensing unit. This unit serves to break the circuit directly or indirectly. In a manual starter, an overload trips a mechanical latch, causing the starter contacts to open and disconnect the motor from the line. In magnetic starters, an overload opens a set of contacts within the overload relay itself. These contacts are wired in series with the starter coil in the control circuit of the magnetic starter. Breaking the coil circuit causes the starter contacts to open, disconnecting the motor from the line.

There are two classifications for overload relays—*thermal* and *magnetic*. Magnetic relays react only to current excesses and are not affected by temperature. Thermal relays can be further divided into two types—*melting alloy* and *bimetallic*.

Fig. 12–4 shows the location of the overload in the circuit with the motor windings. In this case it is a two-terminal external overload device. There is a run capacitor but no start capacitor or relay in this circuit. Fig. 12–5 shows a motor overload device firmly attached to the

Fig. 12–2. Screw-in type of fuse (top) and one-time knife type of fuse.

Fig. 12–3. Cutaway view of a dual-element fuse.

motor housing. It quickly senses any unusual temperature rise or excess current draw. The bimetallic disc reacts to excess temperature and/or excess current draw and flexes downward, thereby disconnecting the motor windings from the power source.

Fig. 12–6 shows how the overload device is connected in a motor control circuit with a potential relay to remove the start winding from the circuit once the motor has come up to speed. Note the two capacitors.

Most small motors are protected from overheating by a bimetallic disc that has a red button sticking out of the motor frame. Fig. 12–7 shows how the circuit breaker operates inside the motor. It stays off until the red button is pressed. If the motor is still too hot, it will open again. It usually takes a couple of minutes to cool off before being able to complete the circuit again.

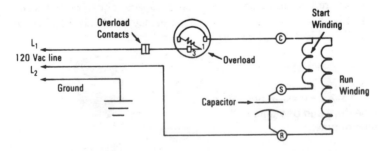

Fig. 12–4. Two-terminal external overload with run capacitor but no start capacitor or relay.

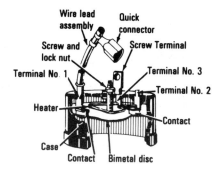

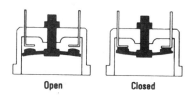

Open   Closed

Fig. 12–5.   External line break overload.

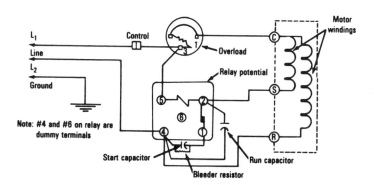

Fig. 12–6.   **Two-terminal external overload with start components and potential relay to remove the start winding from the circuit.**

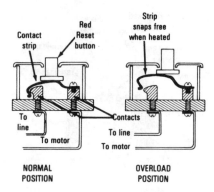

Fig. 12–7. Push-button reset. Located on most small-horsepower motors. The button is usually red and mounted in the motor frame.

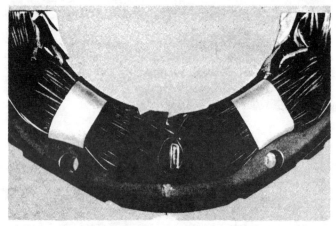

A thermal protective device guards motor windings and insulation from damaging heat which can seriously decrease motor life expectancy. The thermally protected motor has an automatic reset device, connected in series with the power line, which acts to break the circuit when overheating occurs in the motor. Possible causes of overheating may be rotor jamming or loss of normal cooling air. The protector is "nested" for firm contact with the winding to insure uniform protector operation.

Fig. 12–8. Internal line break overload. This is wedged inside the motor windings. It automatically resets when the windings cool down.

Internal (Line Break) Motor Protector

Another safety feature for a motor is the internal line break over-load, which is completely internal and tamper-proof; it cannot be by-passed. This type of circuit breaker is located precisely in the center of the "heat sink" portion of the motor windings. It protects the motor by detecting excessive motor winding temperature and protects the circuit source of power as well (Fig. 12–8).

This type of circuit breaker is wired in series with the contactor holding coil and the pilot circuit contacts of the external supplementary current-sensitive overloads.

If the motor temperature rises above safe limits, the thermostat opens the holding coil circuit. The contactor disconnects the motor from the power source. Supplementary overloads do not break line current but instead interrupt current flow to the contactor holding coil. The contactor then disconnects the motor from the power source (Fig. 12–9).

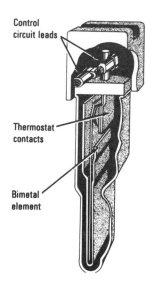

Control
circuit leads

Thermostat
contacts

Bimetal
element

Fig. 12–9. Internal pilot circuit thermostat and supplementary overload embedded in motor winding.

Fig. 12–10. Enclosed switch with handle guard and pilot light. *(Courtesy Square D)*

Manual Starters

Fractional-horsepower manual starters are designed to control and provide overload protection for motors of 1 horsepower or less on 115 or 230 volts single-phase. They are available in single- and two-pole versions and are operated by a toggle handle on the front.

Manual motor starting switches provide on-off control of single-phase or three-phase AC motors where overload protection is not required or is separately provided (Fig. 12–10).

Motor Controls

Fig. 12–11 illustrates some of the various types of switches used for motor control. These are very much a part of the control circuits for

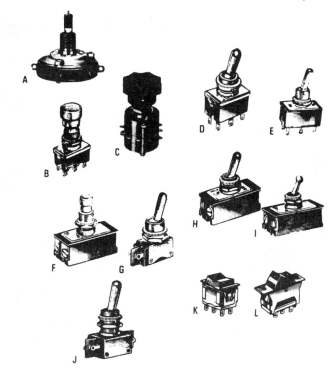

Fig. 12–11. Different types of switches used for motor control.

fractional-horsepower motors. The toggle switches have a pole sticking up so they can be operated in the on-off position by flipping the switches. A few of the different types of switches used to control motors are explained: Switches A, B, C, F, G, and H are toggle switches. Switches D and E are rocker types. It takes a rocking action on the part of the switch control to operate the on-off portion. Switches I and J are push-button types. They can be either momentarily on and then off or they can be pushed for on and pushed for off. Switches K and L are the rotating type. They are rotated to on and off positions.

Maintenance and Repair of Switches. In most cases it is best to remove the defective switch and replace it with an equal or better type. Make sure you check to see if the switch will operate with an inductive load. The motor presents an arc once the switch is turned off and the magnetic field of the motor coils collapse. This means that the motor control switch must be able to take a greater current surge than a switch you'd use for turning on and off a regular light bulb or resistive load.

The quickest way to check a switch is to remove it from the circuit and test it with a continuity checker or with an ohmmeter. When the switch is off, it should read "infinity." When the switch is on, it should read a direct short on the ohmmeter. Any deviation from this indicates the contacts of the switch are not working properly. It is very difficult to test for proper operation of a switch with an ohmmeter and be 100 percent sure the switch is good. The low voltage of the ohmmeter does not indicate that the switch is being tested under the same conditions as when it is in the circuit with power applied at its full voltage. This check with an ohmmeter will, however, give you some indication of good or bad. After checking the power source to make sure power is available and checking the fuse or circuit breaker on the motor, the next thing to check when a motor does not operate is the switch. If you have the proper insulated alligator clips on the ends of a piece of jumper wire, you can short across the switch contacts to see if the switch is operational. If the switch is defective, it will not operate the motor. However, if the jumper is placed across the contacts of the switch and the motor operates normally, it indicates that the switch is defective. *Replace it!*

High-Resistance Contacts

Some switches develop high-resistance contacts after prolonged use. This means they may draw too much current at the contact point and arc, thus creating a hot switch. If the switch is hot to the touch, it should be checked to make sure that the line voltage is reaching the motor terminals. Use your voltmeter to check the line voltage and then check the voltage available at the motor terminals.

If the voltage isn't the same at both the checked points, you should replace the switch with a new one of equal or better current ratings. Also, make sure the voltage rating of the switch is sufficient for the motor circuit.

Microswitches

Microswitches are used in many switch devices. They are usually enclosed in a box or some other type of enclosure. There may be any

Fig. 12–12. Microswitches. Note the different methods by which the switch can be actuated.

number of ways to operate the switch contacts. Levers, rollers, and direct force are used to operate the contacts inside the microswitch. The dimensions on the units will give you some idea as to size. These small switch contacts can handle large currents (Fig. 12–12).

When checking a microswitch, look for NO, C, and NC marked on the side. The C is "common." NO is a "normally open" contact that will read "infinity" on an ohmmeter when checked between NO and C. When checked with an ohmmeter, it should read "zero" on NC ("normally closed") and C. Now, if you press the lever, the contacts will switch their arrangement—that is, C to NO will read "zero," and C to NC will read "infinity." Of course, some microswitches have but two contacts. This means that the switch is normally open if not otherwise stated.

Relays for Motor Control

With manual control, the starter must be mounted so that it is easily accessible to the operator. With magnetic control, the push-button stations or other pilot devices can be mounted anywhere on the machine, and connected by control wiring into the coil circuit of the remotely mounted starter.

Armature

In the construction of a magnetic controller, the armature is mechanically connected to a set of contacts, so that when the armature moves to its closed position, the contacts also close. The drawings in Fig. 12–13 show several magnetic and armature assemblies in elementary form. Fig. 12–14 shows what a motor control relay looks like in operational form. Fig. 12–15 shows the principle on which the magnetic controller (relay) works. The electromagnet makes the difference

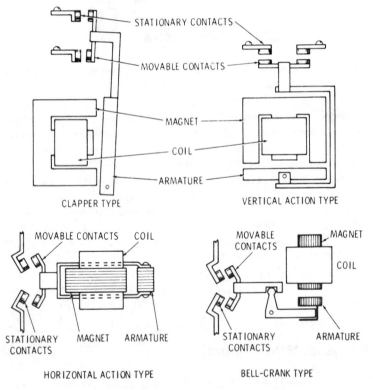

Fig. 12–13. Magnetic frame and armature assemblies.

Relay contacts

Fig. 12–14. Relay. Double pole-double throw.

between a remote control possibility and a manual starter control. The electromagnet consists of a coil of wire placed on an iron core. When current flows through the coil, the iron bar, called the armature is attracted by the magnetic field created by the current in the coil. To this extent, both will attract the iron bar (the arms of the core). The electromagnet can be compared to the permanent magnet shown in Fig. 12–15.

The field of the permanent magnet, however, will hold the armature against the pole faces of the magnet indefinitely. The armature could not be dropped out except by physically pulling it away. In the electromagnet, interrupting the current flow through the coil of wire causes the armature to drop out due to the presence of an air gap in the magnetic circuit.

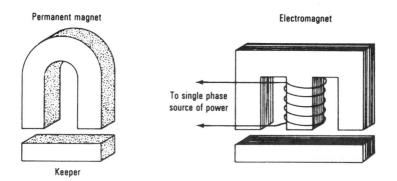

Permanent magnet

Electromagnet

To single phase
source of power

Keeper

Fig. 12–15. Basic principles of a relay operation.

In the other part of Fig. 12–15, the coil, made up of the wire wound around the core, is controlled by a switch located near the power source and any distance from the electromagnet switch it is part of. The contacts can be attached as shown in Fig. 12–13. Once the coil is energized by closing a switch at some location and completing the circuit to the coil, it attracts the armature, which has the contacts attached to it. Once the armature is near the iron core of the coil, it closes the contacts, completing the circuit. The contacts are the switch points which complete the circuit to the motor coils.

Maintenance and Repair of Relays

If the relay contacts do not close, a number of things can cause the condition. If the contacts do not close and there is no magnetic field around the core of the electromagnet, then there is no power being applied to the coil. Check to see if you have power. This means checking the fuses and circuit breakers associated with the line current. If you manually (carefully, with an insulated piece of fiber or similar material) close the contacts and the motor starts and runs while you have the contacts closed, it means that the coil or the switch controlling the coil circuit could be defective. Remove all power and check the resistance of the coil. It should give you a reading on the ohmmeter. If it reads "zero," it is shorted and useless. If it reads "infinity," it is open and must also be replaced. If you get the proper reading (this will vary with the manufacturer), it means the coil is okay but the contacts may not close when the coil is energized by pushing its control switch. Check the switch in the control part of the coil circuit. If the switch is okay and the coil produces a hum and will not pull in, check for low voltage to the coil. Use your voltmeter across the terminals of the coil.

Contacts

If the contacts of the relay are pitted or burned, it may mean you need to burnish (polish) them so they will close tightly against one another and complete the circuit. This can be done (with the power off, of course) by placing a piece of sandpaper (very fine grain) between the contacts and holding the contacts closed. Then move the sandpaper till it sands down the high points on the contacts. Use an even finer grade

Fig. 12–16. Relay contact burnisher also used for cleaning switch contacts for start windings in split-phase and capacitor-start motors.

and polish the points further. Make sure you don't get the contacts to the point that they do not "mate" properly. It may be sufficient to use a tool such as that shown in Fig. 12–16.

Contact Burnisher

The relay contact burnisher is used to polish contacts on switches and relays. It looks like a large ball-point pen. The cap comes off to reveal a small, very thin piece of serrated metal (file). The file is used to get between the contact points and clean them till they are polished. Different degrees of surface roughness are available for the tool. Just make sure you don't try to polish relay contacts when the power is on.

Maintenance of a relay means checking it occasionally for the condition of the contacts and polishing them for good contact. It also means keeping the contacts free of dirt, dust, and, in some cases, extreme humidity and water. It is a good idea to enclose the relay in a waterproof container. Many types are made; they are available at any electrical supply store.

One thing is very important in replacing a relay. Make sure the coil voltage is the same as the one you're replacing. There are hundreds of voltage possibilities in relays. Also make sure the relay contacts will handle the motor current.

Electronic Speed Control

The universal series-wound motor is easily controlled by a circuit such as shown in Figs. 12–17 and 12–18. The SCR (silicon-controlled rectifier) is the heart of the control circuit. It is possible to obtain low speed and good torque with this type of control. Usually this type of circuitry (or something similar) is used to control the speed of a hand drill or other small appliance such as a kitchen blender or mixer. There

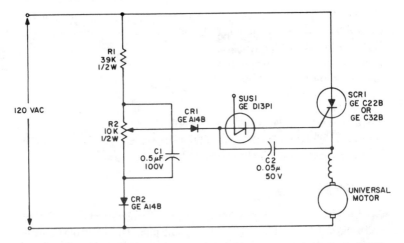

Fig. 12–17. Triggered universal-series motor-speed control circuit with feedback. *(Courtesy General Electric)*

is a smoother change of speed than with switches moving from one position to another. In this type of circuitry, the volume-control type of resistor (R_1) shown here is used to actually control the amount of current allowed to flow through the armature and field coil of the motor.

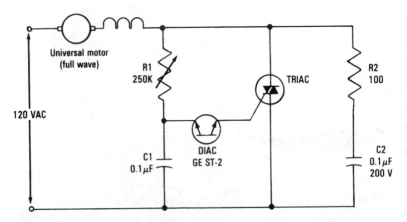

Fig. 12–18. Another type of universal motor-speed control. *(Courtesy General Electric)*

If the motor reaches full speed when the full position is reached on the control but it does not run when the other positions are in the circuit, then it is time to replace the entire electronic unit unless you have the proper experience and equipment to check the SCR and associated capacitors in the circuit. If the speed control works and causes the motor to vary in speed but not reach full speed, then the problem is the switch. This can, in most cases, be removed and replaced to make the control operate properly again. In most cases of this type, the speed control is located inside the cover of the motor or appliance. Care must be exercised in removing the control section.

Control Box

In some instances you may want to control the speed of a motor you already have, but it does not have control built in. This you can do with a control box (Fig. 12–19). The SCR and its associated circuit parts are encased in a box where the power line comes out one end and the AC socket is adjacent to it. Just plug the box into the wall socket and plug the motor into the AC outlet. The control knob will vary the speed of the motor as it is rotated from "0" to the "Full" position. At

Fig. 12–19. Packaged speed control for universal and DC motors.

Fig. 12–20. Foot-controlled speed control for a sewing machine or DC motor.

"Full" position it should place the full line voltage across the coils and armature of the motor and full speed should be obtained.

Keep in mind that the speed control of this type is usable only on universal motors and DC motors. This type of control cannot be used with split-phase or capacitor-start motors. The SCR actually changes the AC line voltage to DC. This DC will work with a motor with a wound armature such as the universal motor. Remember that the universal motor means it will work on AC or DC, but if it doesn't have brushes, it won't work with this type of speed control.

Foot Pedal

Another type of speed control is the foot-pedal control used with sewing machine motors (Fig. 12–20). This is nothing more than a resistor that has its resistance varied when the foot pedal is depressed. One thing to remember with this type of control and universal motors is that a resistor in the circuit causes the universal motor to have next to no torque when the motor is started from zero rpm. That is why most sewing machine motors have to be started by hand each time the machine is stopped and started again.

About the only trouble you have with this control is when it is used on a motor of greater than $\frac{1}{15}$ horsepower. The contacts of the switch become pitted and cause more resistance in the circuit than is called for in proper operation. It can be easily disassembled and checked. Remove the screws in the bottom plate of the control and check for dirt, dust, or any corrosion that may have formed due to humid conditions.

Clean the contact points with sandpaper and blow out the sand dust. Check with an ohmmeter to see if it will vary the resistance once reassembled.

The best care can be that of *not* going from "0" to "Full" in one quick motion. This can create sparking or arcing of the contact surfaces between the slide and the resistance wire. Arcing causes damage to any surface since the temperature approaches 2200°F at the source of an arc.

CHAPTER 13

Armatures — Stripping and Rewinding

Actually, the burnt-out armature is rare in a motor. This does happen, however, in electric hand drills, hand saws, and other small motors that use a wound armature. These small armatures are largely nonrewindable by hand and are usually discarded or replaced. In modern production of motors, the armature is wound by machine. In some instances it is impossible to obtain a replacement armature. Therefore, this chapter will deal primarily with the rewinding of an armature and stripping it before starting the rewinding job. The fractional-horsepower motors with armatures are primarily the small portable equipment types with brushes and commutator, designed to operate on AC.

Most of these armatures are wound with a magnet wire that uses FORMVAR insulation. FORMVAR was introduced in 1938 to replace the double-coated cotton wire. It is polyvinyl formal and has some rather admirable characteristics.

Clear, plain enamel of good quality is still used commercially. Asphaltic enamels are also used in low-cost, disposable fractional-horsepower motors. These coatings for magnet wire have served well for a number of years.

Armatures have also been coated once they were completed and tested. This coating of varnish is what they refer to in the trade as an after-treatment. It has to be removed before the copper can be taken out of the slots and reclaimed. It is important to save the wire that is removed for reuse. This does not mean for you to reuse it in motor rewinding, but turn it in so it can be used in a reclaiming operation. The return from scrap copper can be sizable and can be used to buy needed tools and equipment.

Checking the Armature

One of the first jobs awaiting the armature rewinder is the checking of the armature to make sure it needs rewinding. It is best to make sure it is damaged and needs to be rewound. A number of checks or tests can be run on the armature before it is stripped and rewound.

Check One. One of the first checks to perform is the high-voltage leakage test. This calls for you to place one probe of an insulation tester (megger—shown in Fig. 13–1) on a segment of the armature that has just been removed from the motor and placed on a wooden-surfaced table. Place the other lead from the megger to a metal portion of the armature, either the shaft or the laminated core section. Turn the handle and check the insulation reading. If you get a very low resistance reading, it is finding a path to ground or to the metal part of the armature. Check each commutator segment one at a time with the other lead still attached to the shaft or to the laminated portion of the armature. All of these should give a very high reading. A high reading indicates normal operation, and this type of reading should be expected.

The megger in Fig. 13–1 has the ability to generate up to 2000 volts for testing insulation resistances of 5000 million ohms. That is

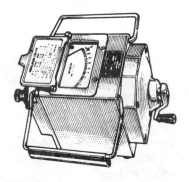

Fig. 13–1. **Megger. So called because it is used to measure extremely large resistances in the millions (meg) of ohms.**

5,000,000,000 ohms resistance. It is also reasonable to assume that if you touch the two probes from the megger and turn the handle for generating the high voltage, you will get a shock. This type of instrument is very important in locating intermittent shorts, bad electrical connections, insulation breakdowns, and conductor failures due to wear, moisture, corrosion, and other causes. It will read out leakage of insulation from 5 ohms to 5000 megohms. It can be used on the bench and in the field since it is fully portable.

Check Two. This test involves testing for shorted windings. Take an ohmmeter and place it on the low-resistance scale—usually the R × 1 scale. Test from one segment of the commutator to another until you get a reading. This should be a complete circuit of a coil. However, it may not be the complete circuit of just one coil. Take a look at Fig. 13–2 to see what is meant by windings. Note how the coils are connected together and then to the commutator segments. Taking the resistance checks will help you locate where the coils are connected and will aid in determining the type of winding used. Keep in mind that the commutator segment has the end of one coil and the beginning of the next coil attached to it.

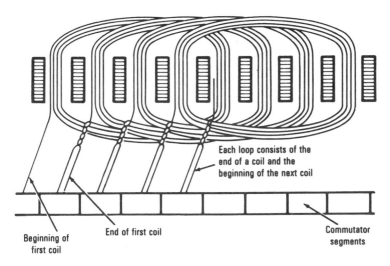

Each loop consists of the end of a coil and the beginning of the next coil

Beginning of first coil

End of first coil

Commutator segments

Fig. 13–2. A lap winding with one coil per slot. This is the type used in universal motors.

Universal Motors

This type of motor can be used on both AC and DC. It has a wound rotor or armature. The universal motor almost always has two poles, which means the windings are all lap windings. The cross-connected commutator cannot be used. The windings are the same as DC motors or repulsion-start motors in principle only. This type of motor uses what is known as the back-lead winding.

The back-lead winding has the wire start at the back of the armature—the end opposite the commutator—and all the connections are brought out as shown in Fig. 13–3.

Burning Out the Varnish

The varnish which insulates the wires will have to be removed. It is difficult to find the proper solvent to dissolve it, so the best way is to use an oven. The temperature of the oven is usually safe at around 350°C. Make sure it is totally enclosed in case there is a fire and the insulation burns or flares up. Gases generated by the burning out of the insulation can be dangerous. *Don't inhale them!* Use the oven in a well-ventilated room. Do not handle the armature until the burn-out has been completed and the insulation is removed. The wire will have a different (dull or black) appearance and will be loosely held in the slots between the laminations.

Data Recording

It is very important to record the wire size that was used in originally winding the armature. This should be recorded on a card and

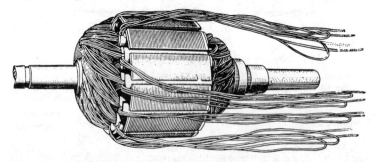

Fig. 13–3. A finished winding of an armature showing the back-leads being brought forward to the commutator.

kept with the armature all during the rebuilding and stripping operations. It is almost impossible to obtain the information once the armature is stripped and ready for rewinding. If the manufacturer's data are not available and technical booklets do not have the information, it is almost impossible to rewind the armature unless you have had years of experience. It is therefore very important to record the data as you strip the armature.

Steps to use in taking the data and stripping the armature:

1. Remove all bands on the armature.
2. Record on your card the type, diameter, and number of turns of the cord bands. Record the diameter and number of turns of the steel wire (if used) and the dimensions and thickness of the clips used to hold the band in place.
3. Take a coil that is exposed across the rear end of the armature from slot to slot and mark it.
4. Remove the wedges or the polyvinyl slot strips that hold the wire in place. This can be done by using a setup similar to the one in Fig. 13–4.
5. Disconnect by cutting the coil lead or leads from the commutator segment to which the coil is attached. In some instances you may have to remove more than one lead from the commutator bar to locate the top coil lead.
6. Check the type of insulation and the wire size. In most instances it is necessary to determine the type of insulation (such as FORMVAR, etc.) before the armature is burned. Record the wire size.
7. Check the top coil's lead by lifting the disconnected leads one by one until one can be lifted easily along its entire length of the slot.
8. Unwind the coil. Each time you unwind a turn, count it. Record the total number of windings.
9. Record the number of bars on the commutator. Each commutator segment is counted as one bar. Count the bars between where the coil started and where it ended. This gives you the commutator *pitch.*
10. The polyester slot liners should have disappeared during the burn-out of the wire insulation. If this has happened, you should check your supplier for the proper size used in an armature of this size. In some instances you would be able to measure the thick-

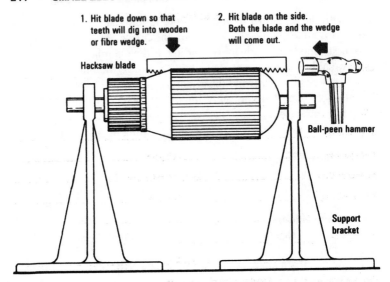

1. Hit blade down so that teeth will dig into wooden or fibre wedge.

2. Hit blade on the side. Both the blade and the wedge will come out.

Hacksaw blade

Ball-peen hammer

Support bracket

Fig. 13–4. Removing wedges from an armature to determine the top lead and where it goes.

ness with a micrometer before the burnout process. If you are stripping by the "brute force" method, the slot liner is still there and can be measured.

11. Make sure you have recorded the type of insulation (usually FORMVAR in small motors of this type) and the diameter of the magnet wire.

Data Card Information

You may want to make up a few cards that will be easily handled in the shop and will give you the informational call-outs as you go along. The card could look like Fig. 13–5.

Slot and Bar Marks

It is important to make sure you have marked the slots and bars. If you have incorrect lead throw, it will cause sparking and poor performance of the motor once it is reassembled. You want to know where the top half and the bottom half of the coil started. You need to know

DATA CARD INFORMATION

MOTOR NUMBER _____ OWNER _____

Horsepower _____ Amperes _____ Volts _____

RPM (r/min) _____ Motor type _____

Serial number _____

Turns per coil _____ Coil pitch _____

Number of slots _____ Motor Manufacturer _____

Kind of winding _____ End room _____

Number of commutator bars _____

Wires per bar _____

Wire size _____ Type of insulation _____

Distance from end of shaft to leading edge of commutator _____

Markings on the shaft or armature core _____

Number of bands _____ Type of bands _____

Insulation for slots _____

Location of bottom lead _____

Additional information _____

_____

_____

_____

Date promised _____ Rewinder _____

Inspector _____

Accepted by Owner _____ Date _____

Fig. 13–5. Data card for recording information needed when you start to rewind an armature.

which slots had the top winding so you can end up that way when you are finished winding the coil. This can also keep you from making a left-hand winding when a right-hand winding is called for. Take a look at Fig. 13–6 to see one of the methods used to mark the slots and the bars.

Markings. Various methods are used for marking the commutator bars and the laminations. You can use a punch and make a (•) or (••) or you can use a file to make an "x." A three-cornered file can be used to make the mark in the laminations. Look at Fig. 13–6 for the proper designation of the two types of windings—lap wound and wave wound.

In most instances you will be winding an armature from a universal-type motor, so it will be a lap winding.

Commutator Removal

In most cases the commutator is pressed onto the shaft. This means that it will be almost impossible to remove it. Therefore it is best to resurface the commutator in a lathe and rebalance it, rather than remove and replace it. If the commutator is such that it needs to be replaced, it will be less expensive to discard the whole armature and forget the problems associated with removing the press-fitted section.

Larger Armatures. On some larger DC machines it may be necessary to remove the commutator. If the winding is deteriorated and brittle, it is advantageous to remove the commutator so that you can obtain the data. On the large armatures, the commutator is built on a strong flange-like support or shell. It is bolted to the armature spider. On small motors, the commutator hub is usually of cast iron (or special steel) and is keyed and pressed onto the shaft.

In order to remove the wiring and obtain the data needed for re-wiring, you should repeat the procedure already given, but you will have to record the data and then remove the space fillers and tapes from behind the commutator risers until the shaft is exposed. Make sure you record the distance of the shaft from the core to the commutator.

Removing the Commutator

You can remove the commutator with a hook-type puller and arbor plates. The procedure for this removal is shown below.

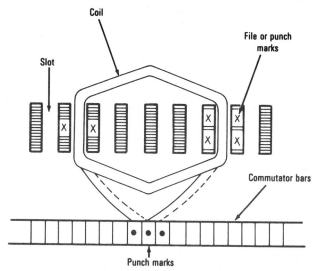

A. Marking of a *lap-wound* armature

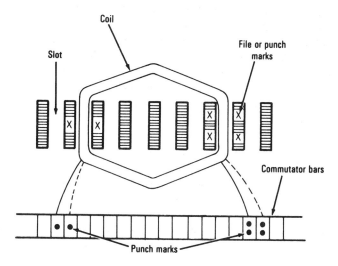

B. Marking of a *wave-wound* armature

Fig. 13–6. Marking the pitch and lead reference points on an armature slot and an armature's commutator segment or bar.

Fig. 13–7 shows the removal of the ball bearing with a hook-type puller. To remove the commutator, place the puller over the commutator and adjust the hooks for the clamping action needed to attach them to the back of the commutator. Adjust the jackscrew to the armature shaft. Slowly tighten the jackscrew with the wrench until the commutator starts to move or slip along the shaft.

Continue the tightening of the jackscrew with the wrench until the commutator drops off the shaft. Place the commutator where it will not be damaged or lost.

If you have not already taken care of the stripping of the insulation from the wire and the armature coils, you can now remove the wedges easily with a hacksaw blade as shown in Fig. 13–4. Since the wedges are wooden on older model motors, they are easily removed with this method. If the newer wedges made of plastic are used, it may take a little more patience.

Other Stripping Methods

Now that you have removed the commutator and the copper portion that may react with certain solutions, you can follow one of two methods of stripping the armature. The idea is to loosen the impregnated varnish. One method is using heat (the oven method has already been discussed), as with an acetylene torch; the other method is to immerse the whole unit into a caustic soda solution.

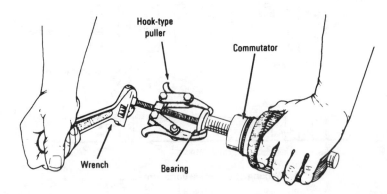

Fig. 13–7. Removing ball bearings from a shaft using a hook-type puller. This method can also be used to remove the commutator.

Using Heat. Remove the slot wedges. Apply heat by using a torch. The acetylene unit should be used the same as for soldering or brazing.

The armature should be mounted between two stands that support it free of burnable materials.

Slowly move the flame so that all parts of the armature are heated as evenly as possible. Concentrate the heat on the core slots and the windings. Don't allow the heat to be applied in any one spot too long. You can damage the shaft if it gets too hot.

By keeping the heat on the windings, you will soon see the windings char and shrink. This will allow the windings to be removed easily. Once the coils and the armature have cooled down, you can then proceed with the removal of the coils of wire.

Using Caustic Soda. Caustic soda solution can be used to loosen the windings of an armature. You have to remove the commutator as previously shown. Then you can place the armature *without the commutator* in a large pan or vat of caustic soda solution. The vat or pan should be deep enough and hold enough solution so that the entire windings are totally covered with the solution.

The pan, without the solution but with the armature, is placed on an electric stove, hot plate, or gas burner. Slowly pour the caustic solution into the vat or pan.

Caution: Wear rubber gloves and handle the caustic soda with care. Do not allow splashing of the caustic soda. It can cause serious burns if it touches your skin. Also use some form of eye protection.

Bring the solution up to the boiling point. Remove from heat and allow the armature to soak until the varnish softens. Remove the armature from the pan or vat once the varnish seems to have loosened. Soak the armature in water and flush it with clean water until the caustic soda solution has been weakened to the point where it is no longer injurious to your skin. Just let the armature drain until it is dry.

Tests on the Commutator

While the armature is drying or cooling, depending on the method used to remove the insulation, you can test the commutator that has been removed from the shaft.

There are a few things which should be done to the commutator before it is checked:

1. Clean between the slots or bars.
2. Use compressed air to remove metal chips and dirt.
3. Make sure that solder has not lodged between bars. If solder is found, remove it.
4. Use an ohmmeter to make checks to ground. You may also use the test rig shown in Fig. 13–8.

Keep in mind that these checks assume that you have already placed the commutator back on the shaft. This way the checks use the shaft as the ground or the common point of reference.

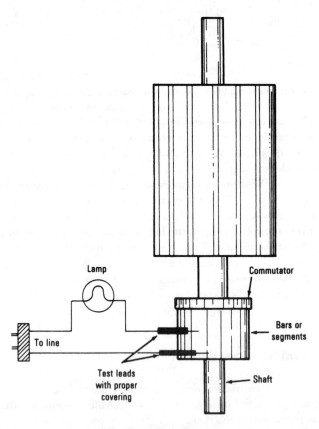

Lamp

Commutator

Bars or segments

To line

Test leads with proper covering

Shaft

Fig. 13–8. Checking the armature for shorts using a trouble light connected in series with a 120-volt source. Be careful of touching the two leads and completing the path for the 120 volts.

If you use the light bulb tester you can:

1. Connect one lead of the tester to the commutator sleeve and touch the other lead to a segment or bar of the commutator. If the light glows brightly, it means that there is a short between that segment and the shaft. Check the commutator to locate where the short is and what is causing it. Remove the cause before going further with the testing procedure. Check every bar.

2. Check for a short from bar to bar after you have finished checking each segment to the shaft. This means using the test light the way it is shown in Fig. 13–9. If the light glows when you are checking *adjacent* bars, it means they are touching somewhere. Locate the source of the short and remove it. Once you have checked the commutator for shorts you can store it until you have time to rewind it.

At this point it may be helpful if you double-check the armature for a clean look. It should have all old insulation removed from the slots. The conductors should be totally removed from the slots—no remnants of the magnet wire should exist anywhere on the armature. Clean the slots with a file if necessary. File off all burrs. Clean the back of the commutator. Clean excess solder and broken pieces of copper out of the bar slots. You probably already did this since you checked the commutator for shorts. If needed, clean out the slots between commutator bars with an old hacksaw blade with the kerf of the blade removed on a grinder. You can use the hacksaw blade to cut slots in the copper part of the commutator to allow you to solder the new wires in place.

Difficult Windings

In some cases it is difficult to remove the windings even after the stripping procedure has been followed and the heat or caustic soda seem to have done the job. You can cut the ends of the windings flush with the core using a hacksaw. Place the armature in a lathe and use a pointed V-shaped tool to remove the end windings at each end of the core. Once this is done, remove the armature from the lathe. Pile up two rows of wooden blocks. They should be high enough to accommo-

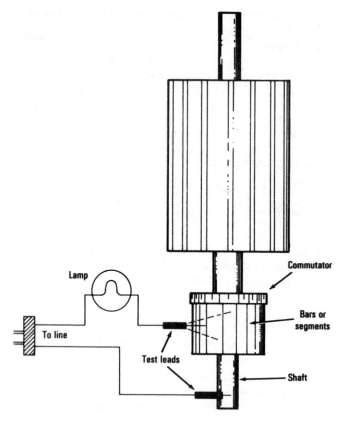

Fig. 13–9.   Checking for bar-to-bar shorts of a commutator.

date the full length of the armature shaft. Rest the core vertically on the wooden blocks. The end of the shaft will just clear the workbench.

Next, drill a hole (larger than the width of the slot) into one set of the wooden blocks. Alight a slot with the hole in the wooden block. This will prevent the outside laminations of the core from being warped when the coil conductors are driven out. You can then use a drift punch and hammer to drive the coil conductors from the slots. Just follow the same procedure for each slot until the wire is removed from the slots. Be careful to check the drilled hole occasionally to make sure it is clear of the coil conductors. This will provide clearance for successive coils when they are driven from their slots. Keep at it until all the old insulation is also removed. Clean out the slots with a file. Remove all burrs.

At this point the armature should be clean and ready for rewinding. Place it aside until you have the time to do the job right. It will take some time for the first job. You will improve with practice and your time for each coil and its proper anchoring to the commutator will decrease with experience.

Rewinding an Armature

Insulating an armature core preparatory to rewinding is one of the most important steps in the rewinding operation. Selection of the proper insulating materials, the method of application of this material, and the proper alignment of core laminations ensure the armature against premature insulation troubles. The following procedures aid in covering all the important points of preparation, selection, and the application of insulation materials.

Problems with Shafts

The most frequent cause of motor failure is bearing failure. If this is not caught in time, there will be shaft failure.

Shaft repair or replacement is not always possible. When some go bad, the whole armature or rotor must be replaced. Keep this in mind when you are inspecting the armature to see if it is worthwhile to be rewired.

Cleaning the Slots and Truing the Laminations

One of the first things to do after all the windings have been cleared from the armature is to clean the core slots, the core ends, and the shaft. All insulation and varnish must be removed. Do this by inspecting the core for rough or irregular spots. Look for sharp edges and burrs. Check for bent, broken, or burred lamination teeth. Remove clinging particles of any old insulation and varnish. Use a solution of 25 percent alcohol and 75 percent benzol. This will help remove stubborn particles. Do not apply any solution to the commutator. Use a cloth moistened, but not wet, with solution for cleaning the commutator.

Use a hammer and drift for straightening the bent slot teeth and laminations. Use an arbor press to force flared or separated laminations together. A file can be used to smooth sharp edges and burrs.

Wedge and Band Grooves

Use a file to clear and true up wedge grooves and band grooves. Place only enough pressure on the file to clear and align the grooves to their original dimensions. If you remove more metal than is necessary, you will cause the wedges and the bands to come loose after installation. Compressed air can be used to blow out loose particles of metal, insulation, and varnish.

Insulating the Core Ends and Shaft

Fiber washers about $\frac{1}{16}$ inch (1.6 mm) thick are used for insulation of the core ends. They are punched to conform to the core-laminations slots. The washers have a tight-fitting hole for the shaft. This serves as a protection for the windings that cross the ends of the core. Reinsulate the core and shaft by cutting the two discs of $\frac{1}{16}$-inch (1.6-mm) fiber to the diameter of the core to be reinsulated. Punch out a tight-fitting hole for the shaft on both ends. Punch out slots to conform to the width and shape of the core slots on both discs. Shellac both ends of the core. Then shellac one side of each fiber disc. After that, assemble and align both end washers to the core. Then you can wrap, half-lap, and shellac about four layers of treated cloth tape (black or yellow cambric) $\frac{3}{4}$ inch (19.1 mm) wide on the armature shaft. Other types are now available. If this is not available, use electrician's plastic tape. Start close to the rear-end washer at a point where the end windings extend. Use masking tape to hold the wrapping down. Tape facilitates rewinding (Fig. 13–10).

Repeat the preceding procedure. Start close to the front-end washer and wrap up close to the back of the commutator.

Replacing the Commutator

The armature of a small motor necessitates the use of an arbor press and properly fitting arbor sleeves for the shaft and commutator.

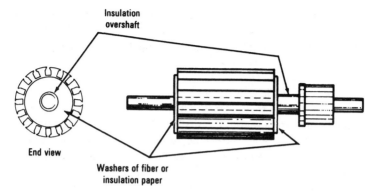

Fig. 13–10. Positioning of washers of fiber over the shaft and against the core assembly before starting to rewind the armature.

In the absence of an arbor press, a correctly-sized arbor sleeve and hammer may be used. The procedure outlined below gives steps to follow when replacing a commutator. Fig. 13–11 shows how the commutator fits snugly over the shaft. In most cases the commutator is shrunk-fit onto the shaft. Be careful when removing it; it is extremely delicate. Spacing between the commutator and the core was determined earlier when you checked the dimensions of the armature before stripping it. Refer to your data card for proper spacing.

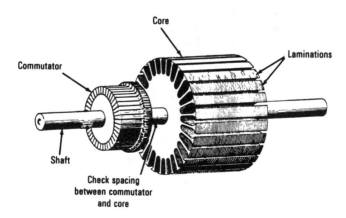

Fig. 13–11.   Stripped armature.

1. Prepare a piece of pipe of sufficient length and diameter for use as an arbor sleeve (Fig. 13–12). Do not use a pipe with an outside diameter that will bear on the commutator bars. The pipe should bear only on the cast-iron hub of the commutator. You can cause irreparable damage to the commutator if you aren't careful.

2. Determine the length of pipe by measuring the distance along the front end of the shaft. Measure from the front end of the armature core to the front end of the shaft. Then subtract the commutator length from this measurement. Recorded data give the distance between the back of the commutator and the front end of the armature core. This should be the proper length.

3. Align the commutator with the shaft. Also align the key and keyway of the commutator hub.

4. Slip the arbor sleeve (pipe) over the shaft. Then align it with the hub of the commutator.

5. Tap the end of the pipe with a hammer. This will get the commutator started on the shaft shoulder.

6. Set the back end of the armature shaft on a block of soft iron,

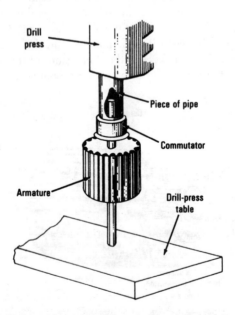

Fig. 13–12. Placing the commutator back on the shaft.

bronze, or brass. It should be ¼ to ½ inch (6.4 to 12.7 mm) thick to fit in an arbor press (a drill press will do fine).

7. Center the ram of the arbor press (or Jacob's chuck of the drill press) on the pipe that is centered on the hub of the commutator (or it may be located on the front end of the armature shaft).

8. Apply gradual pressure until the commutator is driven into the proper position. Be certain that the commutator is driven evenly on the shaft. Do this by constantly watching as pressure is applied.

After the commutator is replaced on the shaft, fill the hollow area between the core and the back of the commutator. Fill the space before or after winding, or fill part of it before and part of it after winding. It will depend on the armature size. The procedure for filling the space follows.

When the bottom of the core slots are lower than the bottom of the commutator riser slots, use a wide tape and wrap it around. Wrap and fill the space to the bottom level of the core slots. Secure the wrapping with a layer of masking tape in case cambric was used to fill the space. Insulate the area and rewind the armature. After rewinding the armature, fill the space between the end windings and the bottom level of the commutator-riser slots with cambric or cotton tape. Secure the wrapping with a layer of masking tape.

When the bottom of the core slots are on a level with, or fairly close to, the bottom level of the commutator, use cambric or cotton tape. Wrap and fill the space to the bottom level of the commutator-riser slots. Secure the wrapping with a layer of masking tape. Insulate. Rewind the armature.

Insulating the Slots

Use a polyester (plastic) slot liner to insulate the slots in the armature (Fig. 13–13). Old motors used fish paper. This paper was supplied in sheets, strips, or cuffed strips in various thicknesses. For 120-volt armatures not wider than ⅜ inch (9.5 mm) and wire for the coils no larger than No. 28 AWG, you will find 0.008– to 0.010-inch thickness paper was used. For 240-volt armatures, insulation at least 0.010 inch thick was used if the coil wire was no greater than No. 28 AWG. The insulation thickness depended on the available room in the slot. For slots wider than ⅜ inch (9.5 mm) and coil wire as large or larger than

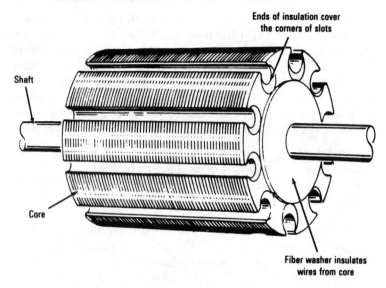

Shaft

Core

Ends of insulation cover
the corners of slots

Fiber washer insulates
wires from core

Fig. 13–13. Bending insulation at slot ends to protect conductor insulation.

No. 18 AWG, they used slot insulation no less than 0.00916 inch thick. Today, preformed polyester slot liners are used.

Use slot liners as insulation between windings (Fig. 13–14). On armatures which operate on voltages higher than 110 to 120 volts, it is necessary to insulate between windings in the slots and between the overlapping end turns of the windings. The procedure (Fig. 13–15) gives the steps to be taken in using the slot liners for insulating windings.

Hand Winding, Coil Forming, and Slot Wedging

It is impractical to use preformed windings on small armatures. End room is limited and the windings must be drawn up tightly to the armature core. In large production shops, small armatures are machine-wound on special equipment designed for the purpose. It is expedient to use preformed coils on large armatures, and this method

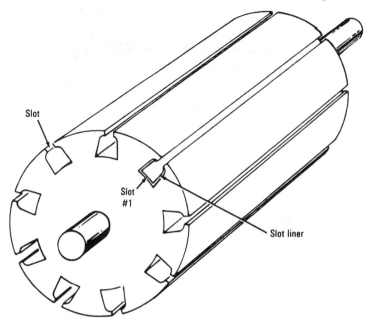

Fig. 13–14. Slot liners (polyester) in place on the armature form.

should be used wherever practicable. Any rewinding job requires the utmost care to prevent the windings from short-circuiting to the metal core.

General Methods of Hand Winding

A hand winding is defined according to the method by which it is wound. There are five general methods of hand winding. They are loop winding, chorded split-loop winding, split-V winding, layer winding, and diametrically split winding. The term *loop* is used to indicate the wire is not cut at the start and finish of each coil. The term *chorded* means that the coil pitch is not full pitch. The advantages and disadvantages of these methods are many. Each has reasons why it should be used and reasons why it should not be used.

Winding the first coil means setting up the reel of wire. Make sure it is of proper size for the armature. Place it on a reel rack with a suit-

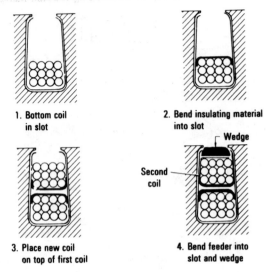

1. Bottom coil
   in slot

2. Bend insulating material
   into slot

3. Place new coil
   on top of first coil

4. Bend feeder into
   slot and wedge

Fig. 13–15. Insulation and slot wedging.

able reel-tension device (Fig. 13–16). Set the reel tension so that little mechanical resistance is offered to a pull on the wire. This resistance causes the reel to stop. It prevents the wire from becoming uncoiled when it is no longer pulled.

For armatures requiring sleeving for lead identification, push lengths of proper sleeving (dependent on lead length) of alternate colors on the wire. The number of sleeve lengths depends on the number of coils for the armature or the number of coils to be wound first. Use different colors for the beginning and ends of each coil. For example, white could be used for the beginning of a coil and blue could be used for the ending. You can improvise a method for holding the sleeving back while you draw the wire for the winding.

A large armature has to be held in an armature stand. However, you can hold a small armature in one hand with the commutator end held toward you (Fig. 13–17).

Insert the wire into a slot (choose any slot and call it No. 1) and hold the end of the wire down on the commutator with your thumb. Wind a few turns into the proper slots. Pull up each turn snugly and firmly. Do not pull so that the wire snaps or cuts through the insulation.

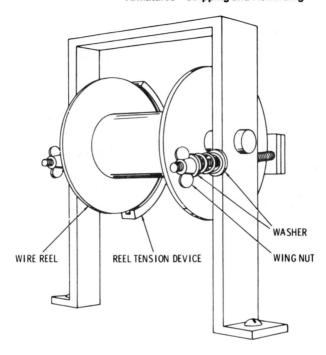

Fig. 13–16. Bench setup of a reel rack.

Release the end of the wire under the thumb and continue winding the coil. Pull each turn snugly and firmly. Do this until the required number of turns is inserted for the winding method you chose for the armature. Pay close attention to the coil pitch.

Press down the windings with a stiff fiber drift plate. This will help get the required number of turns into each slot. Secure the winding in the slot. Exercise care when pushing the coil windings together. Do not damage the winding insulation. Move a length of sleeving to the end of the winding and loop the wire. This is the start of the next coil.

Follow appropriate procedure according to your chosen method of winding for the rest of the coils (Figs. 13–18 and 13–19).

Securing Windings in the Slots. Slot wedges are used for securing windings in the slots. These wedges are of three different types. The use of a particular type depends on the available remaining space in the slot once the winding has been completed. They may be made of

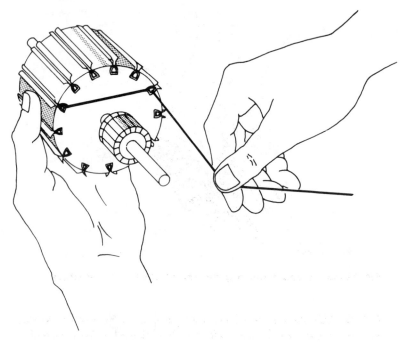

Fig. 13–17. Holding a small armature in one hand during the rewinding procedure. Ends of the coil come out on the other side or toward the back of the armature core.

$\frac{1}{32}$- or $\frac{1}{16}$-inch-thick (0.8 mm or 1.6 mm) fiber or wood; where space is limited, they can be made of polyester.

Wedges for Well-Filled Slots. When the required number of turns are in the slot and a wedge of an appreciable thickness cannot be pushed in from the ends, a polyester wedge, cut just a trifle wider than the slot opening, is used and pressed into the slot from the top. Plastic papers are available in various thicknesses and will easily adapt themselves to the motor specifications.

Put one edge of the wedge under one overhanging edge of the slot. Push the other side of the wedge or paper down under the opposite edge of the slot with a narrow-width steel drift and hammer. An old hacksaw blade with the teeth ground off can be used for a steel drift. This method also applies to open-slot armatures. The wedge makes a

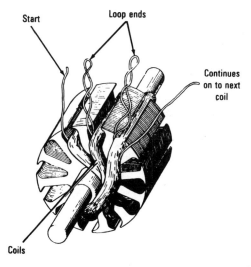

Start

Loop ends

Continues
on to next
coil

Coils

Fig. 13–18. Continuous loop method of winding an armature.

tight fit. After the armature has been dipped in varnish and baked, this type of wedging will hold as well as wedges pushed in from the end.

The selection of fiber or wooden wedges depends on the available room in the slot and on the availability of the materials. Many times it may be necessary to push two wedges either of the same thickness or of different thicknesses into a slot to hold the winding firmly in place. Cut enough wedges of proper length and width for the number of slots in the armature. If long lengths of prepared wedges are available, select the proper width and cut to the proper length. Start a wedge of greater thickness if the one you are pushing is slipping in too easily. The wedge must fit snugly.

Loop Method of Winding

One Coil per Slot

The most popular winding method for small armatures is the loop method (Fig. 13–20). The advantage of this method is that it can be quickly applied and connected when correctly wound. The disadvan-

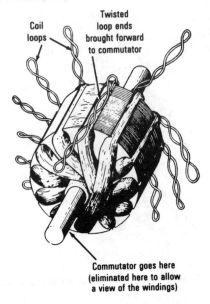

Fig. 13–19. Note how the loop is brought back to the commutator. This has the commutator removed for a better view.

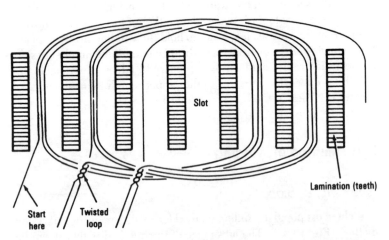

Fig. 13–20. A loop winding showing loops at the end of each coil

tages are that the coils are not all alike in that the first coil is smaller than the last. Also, where there is more than one coil per slot, the resistance of the last coil is greater than the first coil. As a result, the resistance of each coil is different. When the turns per coil are many or the wire size is large, the ends of the coils pile up so that it is hard and sometimes impossible to wind in the coils properly. Due to the unequal size of the coils, a loop winding is not balanced and requires extreme care in balancing after winding.

Loop-Winding Procedure

Use the following procedure for winding armatures with the loop winding (such as universal motors):

1. For armatures which will operate on 110 to 120 volts or higher, place strips of insulation between overlapping end turns of the windings at the back and front of the armature.
2. Wind the second coil into the slot adjacent to the first slot according to the coil pitch. Identify the end of the coil winding with colored sleeving.
3. Make a loop of sufficient length to reach the commutator for connection. Identify the beginning of the third coil with colored sleeving.
4. Wind in the third coil. Follow the same procedure for all the coils of the armature.
5. Cut the wire from the reel when the last turn of the last coil is inserted into the last slot. Connect the end to the single lead of the first coil.
6. Lap windings (shown in Fig. 13–21) can be wound left-hand or right-hand on either side of the shaft without affecting the direction of rotation of the motor or changing the brush polarity of a generator.

More than One Coil per Slot

The winding procedure for this type of loop winding is the same as in the winding procedure for the first coil, except that there will be more than one coil in every slot (Fig. 13–22). For example, a 12-slot armature with 36 commutator bars will require three coils per slot.

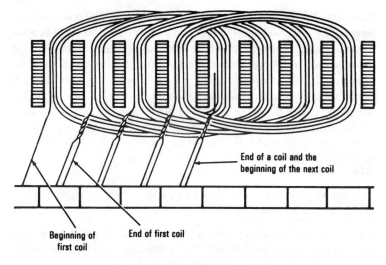

Fig. 13–21. Lap winding with one coil per slot.

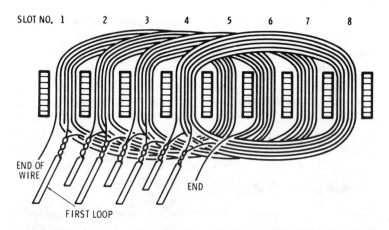

Fig. 13–22. A two-coils-per-slot winding with short and long loops for identification.

Every loop of these three coils must be marked with sleeving of a different color. Red, white, and blue, or any other combination of dissimilar colors, will be satisfactory. Sleeving can be used for one-fourth of the total number of coils of the armature and one-fourth of the total number of coils loop-wound and identified. Winding and identification is continued for the next one-fourth of the total windings. Continue this process until the entire armature is wound. By this method, each slot has three identified loops projecting, except at the end of each one-fourth of the total number of slots, where there would be two identified loops and two identified ends.

Connecting Loops to the Commutator

Tests should be made for open circuits before soldering the leads to the commutator bars. Check for short circuits and grounds. If trouble is found, it should be corrected immediately. When everything is found satisfactory, the armature is prepared for soldering. Do this by wrapping tape around the ends of the leads at the point where they enter the commutator. The commutator connections are then soldered. Use care so that no solder gets on the leads in the rear of the bars.

The commutator is turned and finished with sandpaper *after* completion of the soldering operation. The armature is now ready for banding.

Armature Banding

Since the motors we are working with are small, fractional-horsepower types, steel-wire banding is not necessary. The main thing we are concerned about here is cord banding. This prevents the leads from being pulled loose from their soldered joints by centrifugal force. Some of the universal-type motors reach speeds up to 17,000 rpm. It is quite possible for centrifugal force to cause some damage at this speed.

In cord banding an armature, use enough pressure in winding so that the band will be tight. Use the proper size of banding cord (determine the size by looking at the relative size of the armature). Proper size of banding cord means light cord for small armatures and heavy cord for larger armatures. Refer to steps 1 through 7 in Fig. 13–23 for the correct procedure in wrapping the armature.

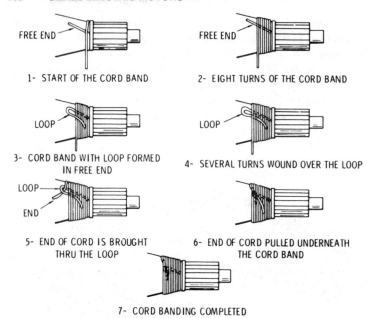

Fig. 13–23. Cord banding of an armature.

1. Start just behind the commutator risers. Allow about 6 inches (152.4 mm) of the free end of the cord to lay at right angles to the circumference of the armature, pointing toward the core slots.
2. Wind about eight turns onto the armature. Each turn goes against

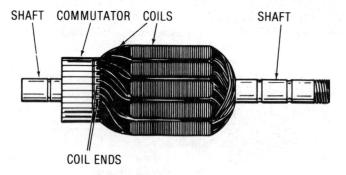

Fig. 13–24. Finished armature.

the other firmly and tightly. Progress toward the core slots and over the free end at right angles to the winding.

3. Loop the free end under the winding and back on itself.
4. Wind more turns (at least as many as put on at first) over the loop ends, each turn against the other firmly and tightly. Progress toward the core slots.
5. Bring the winding end of the cord through the loop. Pull it up tightly.

Fig. 13–25. Armatures for universal motors—machine-wound.

6. Pull the free end of the loop. Bring the winding end under the turns of the cord and out between the turns.

7. Cut the two free ends. Emerge from between the turns close to the turns. Smooth out the turns so that the point where the free ends emerged will be even.

Testing the Armature

It is important to test an armature after winding (Fig. 13–24). Test for shorts, opens, or grounds. If all the precautions are observed during the rewinding procedure, trouble will be minimized.

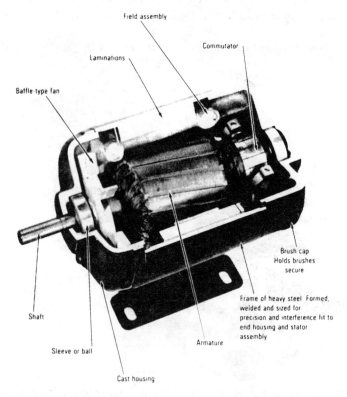

Fig. 13–26.   Armature in place in a universal motor.

Place the armature aside. Begin with the rewinding of the stator, if it needs rewinding. If the stator does not need rewinding, place the armature back into the proper bearings and reassemble the motor. Test for proper operation first by turning the armature by hand and see if it moves freely before power is applied. Check commutator tension. Brushes may be too tight. Apply power. Check for arcing between the brushes and the commutator bars.

Now that you have checked out the armature winding procedure, you will find the stator winding a bit easier. The next chapter will deal with rewinding the stator coils in a motor.

Use a coating of varnish on the entire armature to ensure good insulation. Some recent innovations have made it possible to obtain the insulation materials in a spray can for use by those who are not exactly in need of huge quantities for dipping a coil. Do not coat the commutator bars.

It is also necessary to check for balance. You may have the coils a

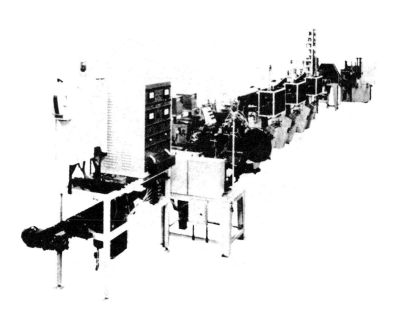

Fig. 13–27.   A semiautomatic armature and motor assembly line. *(Courtesy Mechaneer)*

Winding Patterns

Lead Loop: This pattern is used when it is not practical to hook leads to a tang-type commutator. The leads are pulled into loops for later connection to the commutator. Tooling is available for anchored loops, and lead-on-form tooling is available for pulling lead loops without an anchored lead. The figure illustrates a 2-coil/slot winding pattern; however, 1, 2, or 3 coils/slot can be wound. The varying lead lengths in adjacent coils provide quick identification for commutator hook-up.

Standard Swing Wind: With this pattern, the leads are automatically connected to the commutator tangs. The pattern is generally used on 1 coil/slot armatures, but 2 coils/slot can be wound if there is sufficient distance between the tangs so that adjacent wires do not short. Both anchored and non-anchored lead pulling can be accomplished. The commutator angle can be from 120 degrees to 360 degrees from the coil centers. This provides support for the leads around the armature shaft to prevent wire breakage during subsequent coil winding.

Standard Automotive: This pattern is used primarily on heavy-gauge, low-turn armatures where pressure build-up on succeeding coils does not affect breakage. There must be sufficient room between tangs for the cross over of lead wire. Connection is made to the commutator within the span or dimension of the coil.

Alpha Automotive: When this pattern is used, an alpha loop around the tang separates the lead so that the cross-over shorting is eliminated.

Wrap-Around Alpha: This pattern gives the benefits of the swing wind pattern for reducing strain on the leads, but provides the lead separation by making the alpha connection at the tangs. The tang is positioned diametrically opposite from the coil to provide shaft support for both the exit lead and the entrance lead for the new coil.

Stuff-In-Slot: Slotted commutators are used in high-speed applications where tang-type commutators cannot be used. The stuff-in-slot winding is used with this type of commutator, but precise slotting of the commutator is necessary.

Fig. 13–28. These winding patterns are lap-wound, lap-connected, and can be wound on automatic machines. These are the types of armature windings you will encounter on machine-wound motors.

little lopsided. It may be necessary for you to balance the armature before applying power. Fig. 13–25 shows the number of armatures found in universal motors used for a number of appliances. Fig. 13–26 shows how the armature fits into a motor.

Automated Armature Winding

This assembly line will turn out finished armatures ready for insertion into motor frames, visible on the left of Fig. 13–27. This particular line features semiautomatic staking machines for assembling the shaft,

Fig. 13–29. This machine can wind two-pole and, in some cases, three-pole armatures at high speeds. Winding speeds up to 3500 rpm are available for many-turned, fine-wire applications. Operator action is simple. The operator loads the armature and pushes a cycle start button. The protective hood opens, the operator removes the wound armature, replaces it with a new part, and starts the winding cycle over again. *(Courtesy Mechaneer)*

laminations, insulating tube, and end fibers. A cell machine with swedger for assembling the cell insulators and flaring the ends is the next step in the line. A tang commutator placer, with tang lift station to assemble the commutator, reshape the tangs as needed, and reject those that cannot be reshaped, comes next. The storage unit sticks up on the right end of the line and stores units for winding, thereby controlling the flow of unwound armatures along the line. There are three auto-

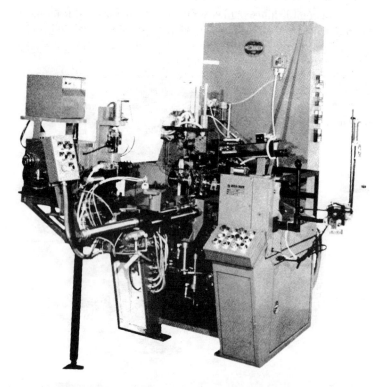

Fig. 13–30. This automatic load and unload winder has an on-board hot staking machine that does not require an operator. The armatures are fed into the machine on an incoming track, automatically loaded into the tooling nest, wound, automatically unloaded into the hot staker for welding the commutator bars, and automatically unloaded from there to a conveyer system taking it to the next operation. *(Courtesy Mechaneer)*

matic winders next. They wind the armature with tang terminations to the start and finish leads.

Next, a storage unit for controlling the flow of wound armatures is inserted in the line. From there a hot stake machine fuses the commutator tangs. The wedge machine assembles the insulators over the windings as the armature progresses along the line. An automatic testing machine makes a complete electrical check of all armatures and includes a reject chute for defective ones.

The staking machine is the only one on the line that is operator controlled. However, automatic stakers are available for a totally automated line. As many as 720 armatures can be handled per hour. That is 12 per minute, or one every five seconds.

Note the winding patterns in Fig. 13–28. These will give you a means of easily recognizing machine-wound armatures.

Figs. 13–29 and 13–30 are armature production machines. They are completely automated.

CHAPTER 14

Stators—Stripping and Rewinding

One of the most important steps in stripping the stator is recording the information you will need later for rewinding. This recording of information consists of taking stock of what is on the stator originally; it means you need to count the number of turns and the size of the wire. Information should be recorded before stripping. Some information is then recorded during the stripping operation. The distance the coils protrude from the stator is important; this is called *end room*. New coils should not extend beyond the slots any farther than this original distance. If they do, the end bells may press against the coils, causing a short circuit. In some cases the end bells may not be able to be replaced at all. A form for recording data for the split-phase motor is shown in Fig. 14–1. Methods of recording the coil pitch data on the form are shown in Fig. 14–2. A typical form for recording data for a polyphase (three-phase) motor is shown in Fig. 14–3. The information must be recorded in such a manner as to enable any repairperson to rewind the motor. Figs. 14–4 and 14–5 show old and new stator windings.

MOTOR DATA CARD

MAKE							
HP		RPM		Volts		Amps	
Hertz		Type		Frame		Style	
Temp		Model		Serial No.		Phase	
No. of poles			End room			No. of slots	
Lead pitch				Commutator pitch			
Wire insulation				Winding			
Slot insulation		Type		Size		Thickness	
Type connections			Switch			Line	

Winding	Type	Size and kind wire	No. of circuits	Coil pitch	Turns
Running					
Starting					

Slot no.	1 2 3 4 5 6 7 8 9 10 11 12 13 14 15 16 17 18 19 20 21 22 23 24 25 26 27 28 29 30 31 32 33 34 35 36 1
Running	
Starting	
Rotation	Clockwise Counterclockwise

Fig. 14–1.   Form for recording split-phase motor data.

Stripping a Winding

Excess-Current Stripping

There are at least three methods used in stripping a stator: excess-current stripping, oven stripping, and lathe stripping. In the first method, the winding can be hooked to a power source. The power source passes current through the winding in excess of what it would normally handle. The heat generated by this method will cause the varnish to melt. The winding then shorts. Place a fuse in the circuit just in case you do not shut off the power at the proper time. Once the varnish is removed, cut the end with a hacksaw or with wire cutters. Remove the wedges. Take a pair of pliers and remove the windings.

SLOT NO. 1 2 3 4 5 6 7 8 9 10 11 12 13 14 15 16 17 18 19 20 21 22 23 24 1

RUNNING

STARTING

COIL PITCH DATA FOR A 24-SLOT, 4-POLE STATOR HAVING OUTER COILS OF ADJACENT COILS IN SAME SLOT.

SLOT NO. 1 2 3 4 5 6 7 8 9 10 11 12 13 14 15 16 17 18 19 20 21 22 23 24 25 26 27 28 29 30 31 32

RUNNING

STARTING

COIL PITCH DATA FOR A 32-SLOT, 4-POLE STATOR.

SLOT NO. 1 2 3 4 5 6 7 8 9 10 11 12 13 14 15 16 17 18 19 20 21 22 23 24 25 26 27 28 29 30 31 32 34 35 36

RUNNING

STARTING

Fig. 14–2. Coil pitch data form.

MAKE	*Westinghouse*			
HP	1/2	RPM 1725	Volts 115/208-230	Amps 8.8/4.4
Hertz 60		Type Cap. start	Frame B 56	Style Open
Temp 40		Model BD 76	Serial No. 312P433	Phase 1
No. of coils 4		No. of slots 36		End room None - Workers
Size of coil		Turns per coil 15		No. of groups 4
Coils/group		No. of Poles 4		Pitch of coil
Size and kind of wire	No. 18 start	No. 24 run	Kind of wire insulation Formvar	
Type of connections:	Switch centrifugal	Line	Thermal protection A	
Slot insulation:	Type Polyester	Size full length	Thickness .003	
Type of winding (hand, machine)	Machine			
Relative position of each winding				
Type of winding circuit	Loop.			

Fig. 14–3. Three-phase motor data form.

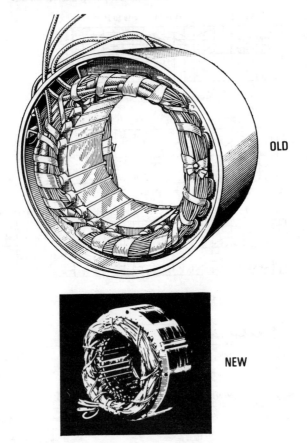

OLD

NEW

Fig. 14–4. Properly taped hand-wound motor stator as it was done in older motors. The other winding is a newer machine-wound type that has been dipped and baked. *(Courtesy General Electric)*

Fig. 14–6 shows how wedges may be removed before you cut the end windings.

Keep in mind that Japan uses polyethylene on its magnet wire. In some cases it is overcoated with nylon for use in submersible pump motors. Magnet wire is usually varnished again for after-treatment or when the winding is finished and some type of binder is needed to hold the wire in place in the slots. You may note also that epoxy varnishes

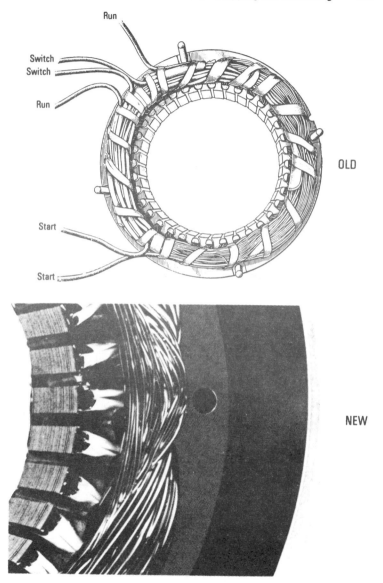

Fig. 14–5. Hand-wound stator (old) and machine-wound (new). *(Courtesy Westinghouse)*

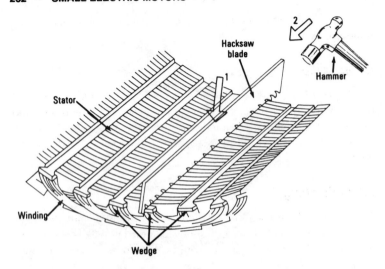

Fig. 14–6.   Using a hacksaw blade to remove wedges.

are popular. Some motor manufacturers use acrylics. Polyester film and integral (fluidized bed) epoxy coatings are used as slot insulators on some motors.

Fig. 14–7 shows how to pull the windings out of the slots with a pair of pliers after the ends have been cut and the varnish removed.

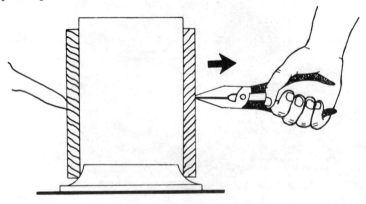

Fig. 14–7.   Use a pair of pliers to pull out the wedges and pieces of wire from the slots.

Fig. 14–8 shows how the motor looks after the windings have been removed.

Oven Stripping

If preheating is done by using an oven, you should cut the coil ends before putting the wound part in the oven. Some methods used by repair shops include placing the motor stator in a hooded area with a gas jet that will burn off the varnish; the hood fan will exhaust the fumes.

Lathe Stripping

You may also use another method which calls for placing the stator in a lathe and cutting the end windings with the lathe tool. This means that a screwdriver can be used to push the windings out of the stator slots afterward.

Two-Pole Motor

If you have a two-pole motor such as is shown in Fig. 14–9, it is an easy matter to remove the coils. The clips that hold the windings in

Fig. 14–8. Stator for a motor (without windings).

Fig. 14–9.   Two-pole motor stator.

place can be slipped out of their seats with a screwdriver. If the coil is hard to remove, use a screwdriver to pry it loose.

Make sure you have checked the wire size and the number of turns before destroying the winding. This one with heavy wire will probably have to be rewound by using a coil form rather than by hand over the slots themselves.

Shaded-Pole Motor

Some types of shaded-pole motors use only a coil wound on a plastic bobbin (Fig. 14–10). The number of turns in the coil may be difficult to determine. However, there are a number of ways to check. If the coil is not damaged, you can use a digital ohmmeter to determine the coil resistance. Then check the size of the wire and match this to the column in wire sizes per square inch and ohms per 1000 feet. Another less reliable method is to check the size of the wire and mark wire limits on the bobbin and set about rewinding the bobbin to the limits of the coil marked. Stripping this type of motor is nothing more than taking the laminations apart. This, in most cases, involves taking the whole motor apart so the coil can be removed, rewound, and replaced. Sometimes you have to grind or drill out the rivets holding the laminations together.

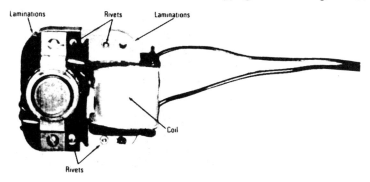

Fig. 14–10. Parts of a shaded-pole timer motor and the location of parts and coil on the motor frame.

Split-Phase Stators

Sometimes it is only necessary to replace the start winding in a split-phase motor. If care is exercised when removing the start winding, the rest of the stator will not need a rewind job. Single-phase motors with more than four poles are a rarity, but if a motor with six or eight poles is encountered, be certain to record the number of poles for the start winding before stripping. They will differ in many instances from that of the run winding. Follow the procedures shown here for both the start and run windings.

1. Remove all wedges. Use a wedge remover or a piece of hacksaw blade with the toothset ground off (Fig. 14–4).
2. Cut through the end turns of the windings. Do this to one end of each pole.
3. Count and record the number of wires entering the slots for every coil side of one side of a pole. The location of the run-winding pole core slots (with respect to the frame) should be recorded by punch-marking the center slot or slots of each pole. In many single-phase stator cores, the center of each main pole is identified by a change in slot size. This is sufficient for properly locating the poles when rewinding.
4. Strip the stator. Do this by using a pair of long-nose pliers on the side of the stator on which the pole windings are not cut (Fig. 14–7). If it is difficult to remove the windings, use procedures for

application of heat or caustic soda solution which are given in Chapter 11. Also, the windings may be connected across full-line current. This causes them to overheat and burn away varnishes and insulations, therefore becoming loose. Repulsion-type motors use the same procedure as for split-phase motors.

Three-Phase Stators

Record the winding circuits for the particular unit being rewound. Do this before stripping. A connection diagram is helpful. After you record the needed data on the stator winding circuits, you can proceed to strip the stator according to the procedure used for split-phase motors (Fig. 14–3).

Rewinding a Stator

The procedure for insulating the slots is similar to that used for the armature. It may be necessary to cut and fit $\frac{1}{16}$-inch (1.6-mm) fiber for the core ends (see the procedure outline in Chapter 13 on armature rewinding). Stator support is furnished by a bench block, its own stator stand, or an armature stand fitted with stator adapters (Fig. 14–11).

Coil Forming

Most single-phase motors are hand-wound. In a very few instances a single-phase stator (composed of preformed coils) may be found. The various methods used to rewind single-phase stators are described below.

Hand Winding. Hand winding is used for both the start and run windings. Wires are placed in the stator slots one turn at a time. This continues until the winding operation is completed (Fig. 14–12). Place slots liners in each slot.

Set up the reel. Choose the proper size wire for the stator and place the wire on a reel rack. The rack should have a tensioning device. De-reel about 10 inches of wire. Place the wire into a slot for the start of an inner coil (Fig. 14–13).

1. Then wind into the slot the proper coil pitch for the inner winding. Check your recorded data.
2. Move to the next larger coil.

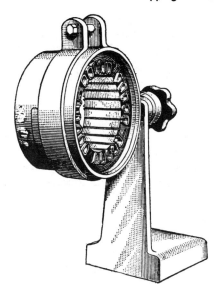

Fig. 14–11. Stator-holding stand.

3. Keep winding the next larger coil until the entire pole is finished. Note how wooden dowels are placed in empty slots. This holds coils in position during winding (Fig. 14–14).

4. Cut the wire. Wedge the winding permanently in place if it is a top winding.

5. Push temporary, loose-fitting wooden or fiber wedges in place for each slot until the stator is completed if it is a bottom winding.

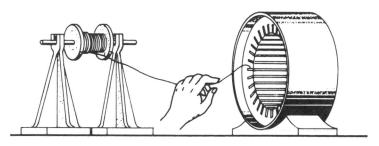

Fig. 14–12. Positioning of the stator and the wire holder.

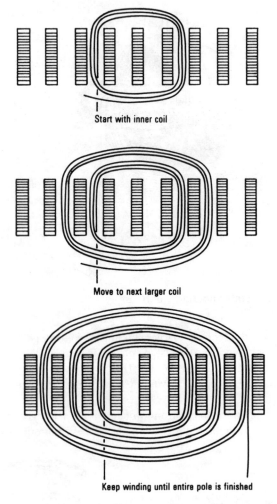

Start with inner coil

Move to next larger coil

Keep winding until entire pole is finished

Fig. 14–13. Procedure for winding a stator pole by hand.

6. Slide the dowels out.
7. Remove the roll of wire. Set up a new reel of wire for the start winding. Make sure it is the proper size. Check your data card.
8. Rewind the start (top) winding after removing the temporary wedges. This is done the same as the run (bottom) winding. Make

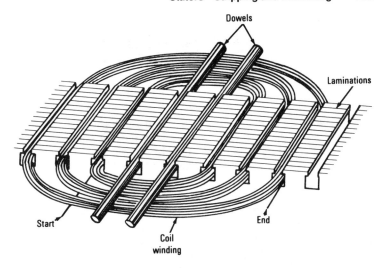

Fig. 14–14. Using dowel rods to hold coils in position while winding.

sure each pole is wound 45° apart from the run winding. This is required for a four-pole stator (90° apart is the rule for a two-pole motor).

Three-Phase Stators

Most three-phase fractional-horsepower motors have diamond-shaped, mush-wound coils. The coils are arranged in lap windings. There are as many coils as there are slots; that means that there are two coil sides per slot. This makes what is known as a two-layer winding. All of the coils are identical in number of turns and size of wire. The phase coils are usually separated by a strip of insulation. A phase coil is a coil adjacent to a coil of a different phase. Coils are wound or connected in groups. A group is simply a given number of individual coils connected in series. Only two leads are brought out. The stator winding is then connected from these groups. In a DC armature lap winding, all the coils are connected in series. In three-phase stators and wound rotors, the coils are connected in a number of groups.

Coils for Small Stators

Wound coils of fractional-horsepower motors are wound with a continuous strand of wire. This simplifies the final connections of a wound stator or rotor. If the number of coils is not the same in all groups, you may want to bring out the leads for each individual coil. Then you can connect them in groups after the stator is wound. Coils for large motors with open-slot stators require a special form and must be wound to the shape of the slot. Their sides must be completely rectangular. Such coils are completely taped.

Single-Phase Motors

Single-phase motors with squirrel-cage rotors have two windings. One is required on the primary to start the motor. This winding is called the start winding and is connected across the line until approximately two-thirds synchronous speed is reached. Then a centrifugally-operated switch automatically opens the circuit.

This start winding is displaced 90° electrically from the main winding. The pole centers of the start winding are located midway between the pole centers of the main winding. The start winding is essentially a resistance winding. It is important that its resistance in the rewound motor be the same as it was originally.

The main and start-winding coils should be connected together as shown in Figs. 14–15 and 14–16. The connections depend on whether the winding is to be connected in series or in series-parallel for the circuit on which the motor is to operate.

Mold and Hand Winding

The mold-winding method is sometimes used to wind small motors of the induction type. The pole coils are first wound on a mold and then placed in the slots. One pole set of coils is wound together so that individual coils do not have to be connected together after being placed in the slots. The mold type of winding has the same general appearance as the hand type.

Any mold-wound motor can be hand-wound. One wire at a time is wound in by hand. When the start winding is mold-wound and no mold

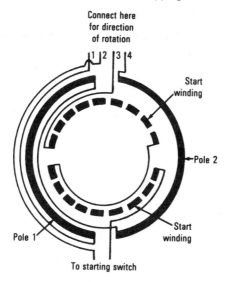

Fig. 14–15. Diagram showing series connections for a two-pole motor.

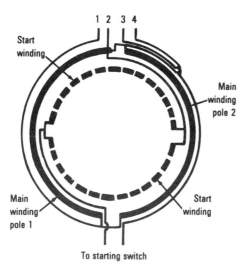

Fig. 14–16. Diagram showing connections for a two-pole, two-path motor.

is available for rewinding, you have to measure the entire length of the start-winding coil of one pole of the motor. Be sure to wind that same amount of wire into the slots.

The split-phase, cage-wound and single-phase, wound-rotor motors most commonly encountered have two, four, or six poles. Two- and six-pole winding distributions and connections are shown in Figs. 14–15 to 14–18.

Internal Switch Connections

Some internal switches have binding screws. If you make connection to these screws, be certain that the end of the conductor is bent around the screw in the direction in which the screw tightens. Make sure the insulation is also removed so that good electrical contact is made between the screw head and the wire. The screw is driven in as tightly as possible without stripping the threads. Other internal switches have soldering lugs or leads. Use the diagram you drew during disassembly of the unit for proper reconnection (Figs. 14–19 and 14–20).

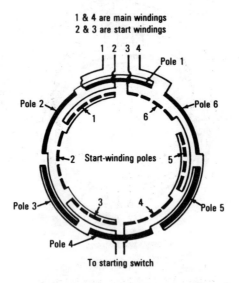

Fig. 14–17. Diagram showing series connection for a six-pole motor.

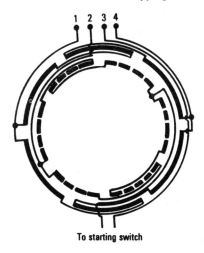

Fig. 14–18. Diagram showing connections for a six-pole, two-path motor.

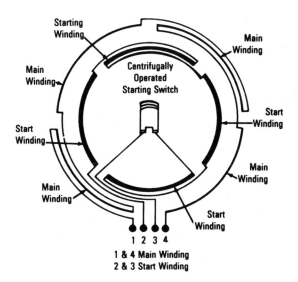

1 & 4 Main Winding
2 & 3 Start Winding

Connect 1 & 2; 3 & 4 for Clockwise Rotation
Connect 1 & 3; 2 & 4 Counterclockwise Rotation

Fig. 14–19. Connection diagram for a series connection of a four-pole motor.

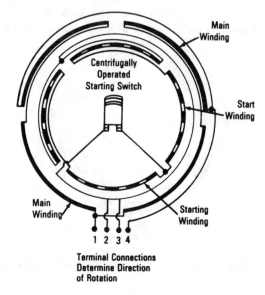

Fig. 14–20. Connection diagram of a four-pole, two-path motor. Clockwise rotation is possible by connecting 1 and 2 to the line and 3 and 4 to the other power line. Counterclockwise: 1 and 3 to one line and 2 and 4 to the other.

Motors That Use Capacitors

There are various types of capacitor-start motors:

Single-voltage, externally reversible.
Single-voltage, nonreversible.
Single-voltage, reversible, with thermostat.
Single-voltage, nonreversible, with magnetic switch.
Two-voltage, reversible.
Two-voltage, nonreversible.
Two-voltage, reversible, with thermostat.
Single-voltage, three-lead, reversible.
Single voltage, instantly reversible.
Two-speed.
Two-speed, two-capacitor.

Keep in mind that motor reversal on split-phase induction motors can be obtained only *by reversing the relationship of the start winding and the run winding.* This may be done externally or internally. The capacitor is placed in series with the start winding at some point of connection.

Electrolytic Capacitors for Motors

Electrolytic capacitors are used in motors. The two-speed, two-capacitor type of motor is really two motors in one. This type of motor has two start windings and two run windings. Each capacitor is connected in series with each of the start windings. A centrifugal switch is common for both windings. Use the diagram developed during disassembly of the motor for reconnection. Check your recorded data.

Testing the Stator

It is important to test a stator after rewinding. Check for shorts, opens, or grounds. Caution in the rewinding procedure minimizes trouble.

Varnishes

Once the motor is tested, it should be varnished and baked. In some instances an air-drying insulating coating is applied over the wires. This can take up to three days to dry.

Electrical varnishes of the oleoresinous species are natural products and are cured by oxidation rather than by heat. They are used for temperatures no higher than 90°C.

Electrical varnishes are used to provide adhesion to increase buildup for insulation purposes and to upgrade overall motor performance. They also protect sensitive elements from environmental contaminants.

Increased buildup over that already on the wire enamel can increase burn-out resistance, heat shock, and thermal rating, and can extend the voltage rating of the motor.

Automatic Coil Winding

Automatic coil-winding machines make the coils to be used in many types of motors. The universal motor that is used in the vacuum cleaner uses a single coil. The shaded-pole motor also uses a single coil. A machine exists that does the winding semiautomatically. The operator has to keep the wire supplied and has to place sealant at the end of the winding so the wire can be cut without unraveling. The operator places the bobbins on the spindle and then starts the machine.

Automatic coil-winding machines come in a number of shapes and forms. The machines wind the armature and the stator, and insert the coils into the stator frame or the armature slots. They also test the finished product and reject it if it does not meet inspection levels.

PART FOUR

Modifications of Motors

CHAPTER 15

Modifying Motors

In order to modify a motor it is necessary to understand its operating characteristics. The characteristics can be changed in certain ways. The method used to change the operational characteristics is chosen with the limitations of the motor in mind. For instance, the most popular electrical types of single-phase motors for general use are capacitor-start and split-phase.

The split-phase motor has two windings, start and run, which are activated to start the motor. The start winding is cut out of the circuit at about 75 percent of operating speed.

The capacitor-start motor is virtually identical to the split-phase motor, but it delivers two to three times the starting torque per ampere of current. This is done by modifying the start-winding circuit and adding a capacitor.

Since the basic difference in these two types of motors is in their load-starting ability, the lower-cost split-phase motor is the logical choice where the starting load is light or where load is applied after the motor has reached operating speed. Capacitor-start motors are required, of course, where heavy loads must be started.

Permanent-split capacitor and shaded-pole motors are designed especially for light starting loads such as time-sequence switches, dampers, and fans.

Characteristics of the different types of motors are shown in Table 15–1. Keep in mind that, as a rule, both motor price and physical size increase as the rated rpm decreases for a given horsepower rating. This means that you may save money by selecting the highest-speed motor that will drive your device.

Motor Modifications for Shafts

A number of variations in adapters for motor shafts are available. This type of modification makes motors usable for different pulley sizes or drive mechanisms. Modification E of Fig. 15–1 shows a shaft adapter for changing a ¼-inch shaft to ⁵⁄₁₆-inch. Modification F shows how to change a ¼-inch shaft by coupling it with an extension 4 inches long. A flat spot is available along the entire 4-inch extension. This will make the motor work in an area where a longer shaft is needed to fit outside an enclosure in most cases.

Modification G changes the ⁵⁄₁₆-inch shaft with a coupling to one with an extra 4 inches. Modification H is a ¼-inch coupling with a ¼-20 threaded extension for a 2-inch addition. The setscrew in the coupler attaches to the existing motor shaft, and the other setscrew in the coupler attaches to the extension so that the ¼-20 thread is used to screw on a pulley or drive device or place the device on the shaft with a ¼-20 nut pulled up tightly to hold it in place on the shaft extension.

A larger shaft coupling and extension is shown in Modification J. This ⁵⁄₁₆-inch coupling has a ⁵⁄₁₆-24 threaded end so the 2-inch extension can be used to handle larger devices. In some instances a shorter extension with screw threads is needed. Modification L shows a hub adapter with ¼-20 threads.

Modification M is a ⁵⁄₁₆-inch coupling and a ⁵⁄₁₆-inch shaft with a 10-24 threaded end that is ⅜ inch long and 6.64 inches long overall. Note the flat spot on one end to fit into the coupling and to be held from slipping on the shaft extension by a setscrew making contact with the flat spot.

Modification N shows a coupling with ¼ inch on one end and ⁵⁄₁₆ inch on the other end and a ¼-inch shaft stock.

These shaft adapters are available from the motor manufacturer or from a local supply house. Many variations can be made in the original motor shafts with these extensions and adapters.

Table 15-1. Typical Characteristics of AC Motors

	H.P. Ratings	Full-Load speeds, rpm (60 Hz)	Starting Torque	Breakdown Torque	Starting Current	Comparative Cost (100=lowest)	Guidelines
Shaded pole	1/65 to 1/20	1650	very low	low	low	100	Low-cost motor for light-duty applications. Compact, rugged, easy to maintain. R&M motors have higher starting torques than most.
Permanent split capacitor	1/50 to 1/3 1/60 to 1/6	3250 1625	low		low	140	Very compact, easy to maintain. High efficiency, high power factor. Can operate at several speeds with simple control devices.
Split phase	1/40 to 1/3 1/50 to 1/6	3450 1725	moderate		high	120	For constant-speed operation, varying loads. Where moderate torques are desirable, may be preferable to more expensive capacitor start.
Capacitor start	1/40 to 1/3 1/50 to 1/6	3450 1725	high		moderate	150	Suitable for constant speed under varying load, high torques, high overload capacity.
Polyphase	1/30 to 1/3 1/75 to 1/6	3450 1725	high		moderate	150	Generally suited to same applications as capacitor-start motors if polyphase power is available. Gets to operating speed smoothly and quickly.

(Courtesy Robbins & Myers)

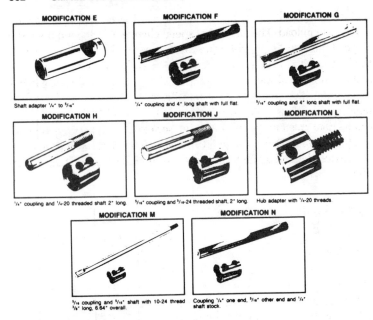

MODIFICATION E — Shaft adapter ¼" to ⁵/₁₆"

MODIFICATION F — ¼" coupling and 4" long shaft with full flat

MODIFICATION G — ⁵/₁₆" coupling and 4" long shaft with full flat

MODIFICATION H — ¼" coupling and ¼-20 threaded shaft 2" long

MODIFICATION J — ⁵/₁₆" coupling and ⁵/₁₆-24 threaded shaft, 2" long

MODIFICATION L — Hub adapter with ¼-20 threads.

MODIFICATION M — ⁵/₁₆" coupling and ⁵/₁₆" shaft with 10-24 thread ⅜" long, 6.64" overall.

MODIFICATION N — Coupling ¼" one end, ⁵/₁₆" other end and ¼" shaft stock.

Fig. 15–1. Modifications on shafts on AC motors. *(Courtesy General Electric)*

Speed Controls

Another way to modify a motor is to change its speed. A number of devices are available for speed changing—in most instances, speed reduction. Both mechanical and electrical modifications can be made to utilize a motor of one speed at another desired speed. However, there are changes in the motor performance which must be taken into consideration when the speed is varied. Different motors react differently. A brief description of some of these differences follows.

Speed Control of Series-Wound Motors

Series motors can be used on AC or DC. They are capable of supplying high starting torque, high speeds, and high outputs.

A series motor inherently does not provide good speed regulation. It is classified as having a varying-speed characteristic. This means

that the speed will decrease with an increase in load and increase with
a decrease in load. The amount of speed change will depend upon the
slope of the motor's speed-torque curve at the load point or throughout
the load range. Furthermore, if a speed control is used, the regulation
and the amount of speed adjustment will be influenced by, or will depend upon, the controlling means. Basically, the speed of a series
motor can be changed by varying the voltage input to the motor. This
can be done in a number of ways. The series motor is very versatile by
virtue of the fact that it can yield a great variety of speed-torque characteristics, depending on its speed-controlling means (the power supply).

Series Resistance Control

Fig. 15–2 shows the resistance method of voltage control, which in
turn affects the speed of the motor. Fig. 15–3 illustrates the principle
using a DC current supplied to the motor by way of a full-wave bridge
rectifier. A variable resistor or rheostat in series with the motor at any
load will decrease the speed as the resistance is increased. Actually the
speed can be adjusted until the motor stops. Due to starting torque
limitations, armature cogging, and reduced ventilation, which causes
overheating, the minimum speed is usually limited to some higher
value.

In the running mode, a series resistor introduces a voltage drop in
the circuit directly proportional to the current. The voltage drop across

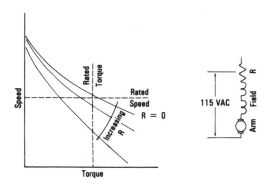

Fig. 15–2. AC series rheostat control. *(Courtesy Bodine)*

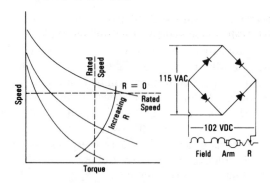

Fig. 15–3. DC series rheostat control. *(Courtesy Bodine)*

the resistor will increase as the motor is loaded since the motor current will increase with the load. It follows, therefore, that the voltage across the motor will decrease with an increase in load and the speed will drop more rapidly with load whenever a series resistor is used. The higher the resistance value, the greater the drop in speed as the load is increased.

When motor speeds are adjusted by resistance, they become much more sensitive to load variations. A given load increase on a series wound motor with resistance in series will cause a larger drop in speed than it would without the added resistance. Also, the value of resistance in a resistor will change with temperature. To maintain speed—as the normal operating temperature is approached—the resistance will usually require adjustment.

The starting torque of a series motor will be noticeably affected by the presence of resistance in series with the winding. This is noticeable especially with a sewing machine, where the motor has to be started by hand each time a universal-type series motor is used to drive the sewing machine. This is due to the fact that at starting maximum current wants to flow, which will limit the motor voltage to its lowest value.

It is not unusual that the motor will not start under full load at the lowest speed or highest resistance setting. The minimum full-load speed at which a series motor will operate with a series resistor is usually limited by the starting torque available to start the load with that value of resistance.

Typically on AC, the speed range of a series motor using a variable series resistor will be from 1.5:1 to 3:1, depending on the motor. On

DC, the speed range will be increased because of the improved regulation and corresponding increase in starting torque.

Shunt Resistance Control Speed

Fig. 15–4 shows this shunt resistance method of speed control using AC. Fig. 15–5 illustrates the principle using DC supplied to the motor by way of full-wave rectifier bridge. For purposes of comparison, Figs. 15–4 and 15–5 were obtained on the same motor used in Figs. 15–2 and 15–3.

The speed range is usually limited by this method because of increased current passing through the field coils and the corresponding heating effect. Care must be taken in choosing the minimum resistance across the armature because the high field current created may burn out the field coils. A wide speed range may be used only if the application has a very limited duty cycle.

Compared to the series resistance method, the shunt resistance method has a narrow or more limited speed range. However, the shunt method of speed improves the speed regulation of the motor and maintains good starting torque characteristics. It is an excellent method for matching motor speeds.

Series and Shunt Resistance

Another method used to modify a motor for speed control is by using a combination of the two systems previously discussed. Varying

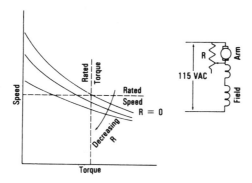

Fig. 15–4. AC shunt rheostat control. *(Courtesy Bodine)*

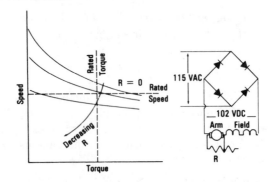

Fig. 15–5. DC shunt rheostat control. *(Courtesy Bodine)*

the rheostats singly or in tandem will give speed-control characteristics of each of these systems as well as those in between. Again, caution must be exercised to avoid overheating the field coils when an armature shunting resistor is used.

Variable Transformer Control

Figs. 15–6 and 15–7 show typical characteristic curves for the transformer control method. The same motor as used in Fig. 15–5 is again employed to make the comparison meaningful.

By using a variable transformer to vary the voltage across a series motor, a speed range of 4:1 to 7:1 is typical, depending on the motor. If a full-wave bridge is used to convert the output of the transformer to

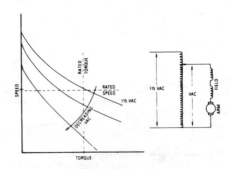

Fig. 15–6. Variable AC voltage control. *(Courtesy Bodine)*

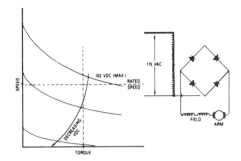

Fig. 15–7. Variable DC voltage control. *(Courtesy Bodine)*

DC, the speed range will be increased because of the improved regulation and starting torque.

Tapped Field Windings

Another way to modify a motor for speed control is by using tapped field windings and a switching arrangement connected to the taps. Such motors provide discrete speed steps rather than a continuously adjustable output.

Governors for Speed Control

To limit the operating speed of a series wound motor to a constant value, governors can be used. Such governors are usually electromechanical devices having a set of contacts that periodically insert a fixed resistor into the circuit to limit the motor speed to a predetermined value.

Electronic Speed Control

Electronic methods of speed control are many and varied. Older methods such as thyratrons or grid-controlled rectifiers, saturable reactors, or magnetic amplifiers have fallen from popularity in favor of solid-state components because of size, weight, and cost. They all have one thing in common—to provide an adjustable voltage to control the speed of the series motor.

Today, a typical electronic series motor control is a half-wave SCR (silicon-controlled rectifier) device with feedback. Since these devices are half-wave, the maximum voltage to the motor is substantially less than 115 volts, resulting in a relatively low top speed from the standard motor used in our comparison in Fig. 15–8. However, the feedback feature—which corrects for drop in speed due to loading—usually allows an extension of the speed range on the low end due to improved starting torque and speed-regulation characteristics.

Prolonged operation of series-wound motors by SCR controls can reduce brush life. Some adjustable-speed series-wound applications obtain best brush life with triac-type electronic controls.

Where excellent speed regulation, wide speed range, and continuous duty operation are required, shunt wound or PM (permanent-magnet) motors with full-wave, feedback-type SCR controllers have become popular in recent years.

For specific applications, motor design factors can sometimes be changed to aid speed-control characteristics. For example, armature slots can be skewed to provide less cogging at the low end of the speed range. This normally will extend the usable speed range.

Extended operation of series-wound motors at low speeds causes overheating unless the motor is specially designed for such operation or external cooling is used.

Many methods are available to modify a motor for the control of speed. The final choice must be made based on the performance de-

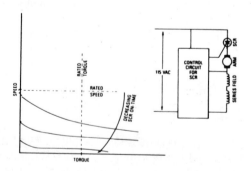

Fig. 15–8. SCR control. *(Courtesy Bodine)*

sired, the temperature limitations of the motor, size restrictions, and overall system cost.

Modifying AC Motors for Speed Control

Modifying AC motors for speed control is not easy. Most of the AC motors are made for a specific speed. The number of poles makes a difference in the speed. The universal motor (AC/DC) is controlled easily by placing any number of speed-control devices in the circuit with the motor. This type of speed control has already been discussed.

AC Motors Designed for Constant Speed

One of the characteristics of the AC induction motor is its ability to maintain a constant speed under normal voltage and load variations. Therefore, as would be expected, this type of motor does not lend itself to a simple method of speed control over a wide range.

There are, however, some variations of conventional induction motors which are designed for the express purpose of improved speed control. These motors may use wound rotors with variable resistance, brush shifting means, and other special modifications.

Keep in mind that the speed of an induction motor is controlled by two factors, the number of poles in the stator winding and the frequency of the power supply.

$$rpm = \frac{F \times 60}{\dfrac{P}{2}}$$

where

$$
\begin{aligned}
rpm &= \text{revolutions per minute} \\
F &= \text{frequency (in Hz)} \\
P &= \text{number of poles}
\end{aligned}
$$

The synchronous AC motor rotates at the exact speed defined by the formula. The nonsynchronous motor never operates at synchronous speed. The difference between the synchronous speed and the actual speed is known as rotor slip.

$$\text{percent of slip} = \frac{\text{synchronous speed} - \text{actual speed}}{\text{synchronous speed}} \times 100$$

The amount of slip depends on the motor design, power input, and motor load. The nonsynchronous motor speed can be adjusted by changing the amount of rotor slip.

AC Motor Speed-Control Methods

Change in Frequency

Change in frequency has the advantage of providing stepless speed changes over a wide range, and may be used with either synchronous or nonsynchronous motors. The major disadvantage encountered with this method is the relatively high cost of frequency of the changing power supply. The frequency-change method must also operate within narrow limits since a motor designed for 50 to 60 Hz usually cannot be satisfactorily operated at a radically different frequency. The motor must be designed for power supply limitations. These may be the laminations and winding design. The inductive reactance of the motor is determined by the frequency and inductance of the winding. The laminations or core and the number of turns determine the inductance and thereby the inductive reactance. The frequency of the power source also determines the inductive reactance. The inductive reactance has a definite bearing on how much the current will be limited. Excess current causes excess heat. This heating effect can take place in a very short time. It is important, then, to make sure the motor is limited in frequency changes. It is difficult to use a 400-Hz motor on a 60-Hz line without excessive current and heat generation being the result. Aircraft use 400-Hz motors, and the 60 Hz of house power will not operate that type of motor since the iron in the core and the number of winding turns are definitely less.

Change in Number of Poles in the Stator

Both synchronous and nonsynchronous motors can use the pole-changing method for speed control (Fig. 15–9). This method of change or modification has a limited number of speeds to offer since the speed is a function of the number of poles. Most motors have no more than

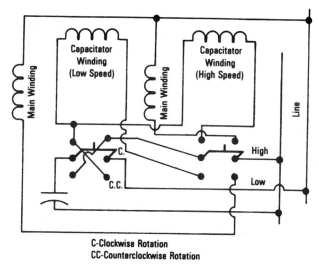

C-Clockwise Rotation
CC-Counterclockwise Rotation

Fig. 15–9. Simplified pole-changing circuit. *(Courtesy Bodine)*

four definite speeds using this method. A portion of the winding is idle during the operation of one or more speeds. This results in motor inefficiency and a considerable reduction in the output rating for any given frame size. Switching methods for pole changing are also expensive and complicated, making the method useful in relatively few applications.

Change in Rotor Slip

The changing of rotor slip is simpler, less costly, and the most widely used technique for varying the speed of an AC induction motor. There are three types of nonsynchronous motors to which this method is best suited. They are the shaded-pole, permanent-split capacitor, and polyphase.

Due to the sensitivity of the centrifugal or relay starting switches, the rotor-slip method should not be applied to split-phase-start and capacitor-start motors unless the speed will never go low enough to engage the starting switch. If the motor is running at the reduced speed with the starting switch closed, the auxiliary winding or the switch contacts will soon burn out.

There are several ways to change the power input to an induction motor. This will increase or decrease the amount of slip. Six modifications are shown here to show how the method works.

Tapped Winding. This method is most widely used for a shaded-pole fan motor. The change in input is obtained by changing the motor impedance through the use of various portions of the total winding (Fig. 15–10). The number of speeds is determined by the number of taps introduced into the winding. In addition to shaded-pole motors, the tapped-winding technique can be used with permanent split-capacitor motors.

Series Resistance. A variable resistor can be used to vary the voltage across the winding of an induction motor (Fig. 15–11). Series resistance can be used with either shaded-pole or permanent split-capacitor motors.

Variable Voltage Transformer. This method may be used in place of a series resistor to reduce voltage across the winding. It has the advantage of maintaining substantially the same voltage under the starting condition when the current is higher than during the running mode. There is also much less power lost as heat than with a resistor (Fig. 15–12). By reducing the voltage across the main winding only of a permanent-split capacitor motor, full voltage is maintained across the capacitor winding, providing more stable operation at lower speeds (Fig. 15–13).

Winding Change. The winding change is applicable to only the permanent-split capacitor motor (Figs. 15–14 and 15–15). The func-

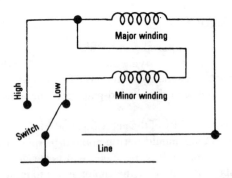

Fig. 15–10. Tapped winding circuit for speed control. *(Courtesy Bodine)*

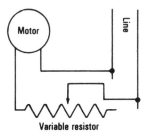

Fig. 15–11. **Simplified series resistance circuit.** *(Courtesy Bodine)*

tions of the main and the capacitor (starting) windings can be switched, to provide a "high" and a "low" speed. High speed is obtained when the winding with fewer turns is functioning as the main, while the lower speed is achieved with the winding with more turns functioning as the main. This is an extremely efficient technique, but it does require that the motor winding be exactly tailored to the load in order to provide the desired two speeds.

Shunt Resistor. Also confined to the permanent-split capacitor motor, this method has been found to provide stable speed in four-pole, 60-Hz motors up to $\frac{1}{100}$ horsepower (7.5 W) over a range of from 1500 rpm down to 900 rpm with constant torque output (Fig. 15–16). With this method it is necessary to use a high-slip type of rotor.

Solid-State Semiconductor. A very popular and practical speed control means in home-installed heating and cooling equipment, semiconductor controls are usually used with induction motors in fractional-horsepower sizes and general-purpose construction using

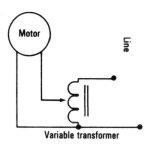

Fig. 15–12. **Variable-voltage transformer method.** *(Courtesy Bodine)*

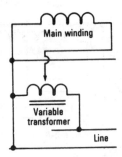

Fig. 15–13. Variable-voltage transformer in a PSC motor circuit. *(Courtesy Bodine)*

standard slip rotors. In the smaller sizes, shaded-pole and permanent-split capacitor motors may be used. For applications requiring larger drives (⅛ horsepower and up), split-phase-start or capacitor-start motors may be specified if the speed range is narrow, with the lowest speed well above the opening point of the starting switch. The high inertial blower-type load typical in heating and cooling equipment plus the relatively narrow speed-range requirement make possible the use of a standard motor and the simple series speed-control circuit.

Fig. 15–14. Permanent-split capacitor winding changes for high and low speeds. *(Courtesy Westinghouse)*

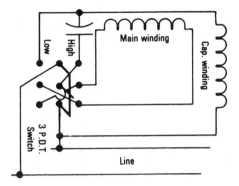

Fig. 15–15. Winding function change method. *(Courtesy Bodine)*

More sophisticated solid-state semiconductor controls have been developed which are capable of satisfactorily controlling the speed of induction motors driving loads with relatively small inertia. Such applications are generally confined to motors in the subfractional range (less than $\frac{1}{20}$ horsepower).

Reversing Direction of Rotation

Some motors come with quick snap connectors that can be switched easily to change the motor rotation. Take a look under the

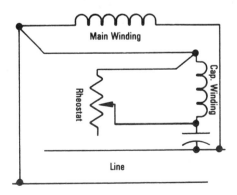

Fig. 15–16. Shunt resistor method.

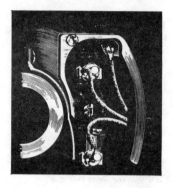 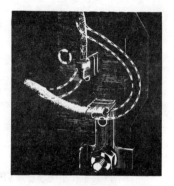

Fast rotation change — just switch two motor leads. Quick connects make changeover fast and positive.

Fig. 15–17. By moving the two leads it is possible to change the motor's direction of rotation. *(Courtesy General Electric)*

removable cover over the box located on the motor. The box for power input is usually located on the side of the motor (Fig. 15–17).

Other motors also have terminals that can be switched easily to change from a 120-volt motor to one that will operate on 240 volts AC. The stator windings are in a series of the 240-volt operation and in parallel for the 120-volt operation. Most motors that have the capability will have the terminals marked on the motor or inside the junction box on the side of the motor.

PART FIVE

Motor Selection and Replacement

CHAPTER 16

Motor Selection and Replacement*

Selection of Motor Type

There are many factors involved in selecting a motor. The application of the motor to a specific job makes it easier. Motors have been designed for many special uses.

Selection of a motor affects its installation; it also affects its operation and its service requirements. Selection may be determined by the user or the motor-driven apparatus requirements. Facts, field tests, and analysis of operating conditions are criteria used for motor selection.

Some basic factors in the selection of a motor are: power supply, horsepower rating, speed, motor type, and the enclosure. Other considerations include motor mounting, motor connection to a load, and mechanical accessories or modifications.

*The authors would like to express appreciation to the Bodine Electric Company and the General Electric Company for assistance in the preparation of this chapter.

Power Supply

Voltage. The system voltage must be known in order to select the proper motor. The motor nameplate will normally be less than the nominal power system voltage. A joint committee of the Edison Electric Institute and NEMA (National Electrical Manufacturers' Association) has recommended standards for both power system voltage and motor nameplate voltages, which are as described in Table 16–1.

Motor Description

Following is a list of the important items making up a motor description:

Design Type—K, KF, P, PF, etc.
Frame—14T, 286T, etc.
Horsepower—through 30 horsepower
Synchronous Speed—1800 rpm, 900 rpm, etc.
Volts—230, 460, 575, etc.
Phases—Single, three.
Frequency—60 Hz, 50 Hz
Enclosure—drip-proof, TEFC, explosion-proof, etc.
Duty—continuous, 1-hour, 15-minute, etc.
Service Factor—1.15, 1.0, etc.

Table 16–1. Power System Voltage Standards

Polyphase 60 Hertz	
Nominal Power System Volts	Motor Nameplate Volts
208	200
240	230
480	440
600	575
Single-Phase 60 Hertz	
Nominal Power System Volts	Motor Nameplate Volts
120	115
240	230

Ambient Temperature—40°C, 65°C, etc.

Mounting Assembly—F-1, F-2, C-2, W-4, etc.

Bearing Requirement—ball, sleeve, etc.

Direction of Rotation—CW (or CCW) facing drive end.

Environmental or Operating Conditions—high ambient temperature, excessive moisture, low voltages, etc.

Special electrical or mechanical features.

Voltage Variation

All motors rated 30 horsepower or less are designed to operate successfully at rated load with a voltage variation of plus or minus 10 percent when rated frequency is supplied. They will also operate successfully when the sum of the voltage and frequency variation does not exceed 10 percent, provided the variation of frequency does not exceed 5 percent above or below nominal ratings as stamped on the nameplate.

Frequency

In addition to operating successfully with a voltage variation, a motor rated 30 horsepower and less will operate successfully with a frequency variation which does not exceed 5 percent above or below its rated frequency.

The predominant frequency in North America is 60 Hz. However, 50-Hz systems are common in foreign countries. Systems such as 25 to 40 Hz are isolated and relatively small in number. However, motors can be designed for these frequencies.

Phases

In general, three-phase power supplies will be found in most industrial locations. However, single-phase only is available for most residential and rural areas. Two-phase power supply is found infrequently.

Motor Design Types

The type of motor will determine the electrical characteristics of the design. The following designs are NEMA designations for three-phase motors.

Design B. A Design B motor is a three-phase, squirrel-cage motor designed to withstand full-voltage starting and developing locked-rotor and breakdown torques adequate for general application, and to have a slip at rated load of less than 5 percent.

Design C. A Design C motor is a three-phase, squirrel-cage motor designed to withstand full-voltage starting, to develop high locked-rotor torque for special high-torque applications, and to have a slip at rated load of less than 5 percent.

Design D. A Design D motor is a three-phase squirrel-cage motor designed to withstand full-voltage starting, to develop 275 percent locked-rotor torque, and to have a slip at rated load of 5 percent or more.

All percentages are in terms of percent of full-load torque with rated voltage and frequency supplied.

Horsepower Requirements

The horsepower required by the driven machine determines the motor rating. Where the load varies with time, a horsepower-versus-time curve will permit determining the peak horsepower required and the calculation of the root-mean-square (RMS) horsepower, indicating the proper motor rating from a heating standpoint. In case of extremely large variations in load, or where shutdown, accelerating, or decelerating periods constitute a large portion of the cycle, the RMS horsepower may not give a true indication of the equivalent continuous load, and the motor manufacturer should be consulted.

Where the load is maintained at a constant value for an extended period (varying from 15 minutes to two hours, depending on size), the horsepower rating will not usually be less than this constant value, regardless of other parts of the cycle.

If the driven machine is to operate at more than one speed, the horsepower required at each speed must be determined.

Motor Mounting

The frame numbers of induction motors specifically identify mounting dimensions. All machines having the same frame designation will have identical essential mounting dimensions, regardless of electrical characteristics, thereby providing interchangeability. For the majority of applications, foot-mounted motors are utilized. The motor is mounted on the driven equipment and secured by four mounting bolts through holes in the feet (Fig. 16–1).

There are a number of methods used to mount motors. The available space and the equipment location make it impossible in some cases to mount the motor on the machine or on the floor next to the machine it is to power. Therefore, it becomes necessary to devise other means of mounting the motors. Some companies will furnish the brackets and others will make the motors according to your specifications if you use enough of them to make it worthwhile (Fig. 16–2). In any case, the following methods of mounting a motor are given to indicate some of the possibilities (Fig. 16–3).

Unless specified otherwise, most motors can be mounted in any position or any angle. However, drip-proof motors must be mounted in the normal horizontal position to meet the enclosure definition. They are mounted securely to the mounting base of the equipment or to a rigid, flat surface, preferably metallic.

For direct-coupled applications, align the shaft and coupling carefully, using shims as required under the motor base. Use a flexible coupling, if possible, but not as a substitute for good alignment practices.

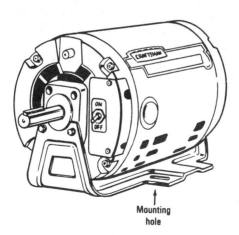

Mounting
hole

Fig. 16–1. **Note the mounting holes in the resilient mounting of the motor.** *(Courtesy Sears)*

Open, Split Phase
Rigid Mounting Base

Open, Capacitor Start
Resilient Mounting Base

Fig. 16–2. The rigid mounting base and the resilient mounting base are shown here for comparison. *(Courtesy Westinghouse)*

For belted applications, align the pulleys and adjust the belt tension so approximately ½ inch of the belt deflection occurs when the thumb force is applied midway between pulleys. With sleeve-bearing motors, position the motor so the belt pull is away from the oil hole in the bearing (approximately under the oiler of the motor).

End Shield Mountings

The industry has standardized the three types of machined end shields which have rabbets and bolt holes for mounting such items as pumps and gearboxes to the motor, or for overhanging the motor on a driven machine.

The *Type C* face end shield provides a male rabbet and tapped holes for mounting bolts. This end shield is used for mounting a small apparatus to the motor.

The *Type D* flange has a male rabbet, with holes in the flange for through bolts. This flange can be used on machine tool gearboxes where the motor is mounted to the apparatus.

The *Type P* base has a female rabbet, with holes for through bolts in the flange, and is used for mounting of vertical motors.

Part-Winding Starting

Part-winding starting is used to reduce the initial inrush of current. Power systems, especially in residential and commercial areas,

Motor Mounting Assembly Symbols
Standard Mounting Configurations

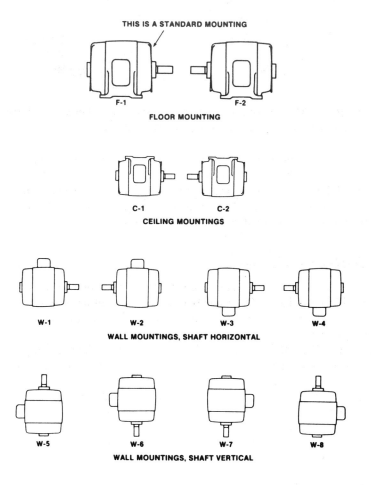

THIS IS A STANDARD MOUNTING

F-1   F-2

FLOOR MOUNTING

C-1   C-2

CEILING MOUNTINGS

W-1   W-2   W-3   W-4

WALL MOUNTINGS, SHAFT HORIZONTAL

W-5   W-6   W-7   W-8

WALL MOUNTINGS, SHAFT VERTICAL

Standard lead location — F-1, W-2, W-3, W-6, W-8, C-2
Lead location opposite standard — F-2, W-1, W-4, W-5, W-7, C-1

Fig. 16–3. Mounting assembly symbols for standard mounting configurations for motors. *(Courtesy Doerr)*

are often limited in capacity. Consequently, when a large motor is started across the line, voltage fluctuation may cause annoying light flicker. To keep this under control, most power companies limit the amount and number of any sudden current demands on their system. Part-winding starting will meet these requirements by putting power into selected portions of the motor windings first and then energizing the rest of the windings a second or two later. The current inrush is limited to about 60 to 70 percent of the normal starting current and the starting torque is reduced to 40 to 50 percent of normal. This reduced current and torque condition exists for only one or two seconds. As the rest of the windings are energized on the second step, another inrush of current occurs, making the total current drawn equal to, or slightly less than, the normal across-the-line inrush current. As all the windings are energized, the motor displays standard characteristics.

Enclosures and Special Winding Treatment

To withstand special or extreme conditions, the following enclosures and/or special winding treatments are available.

Open, Drip-Proof. Normal insulation treatment which consists of one or more dips and bakes of varnish. The insulation system is composed of materials which will not absorb or retain moisture.

Open, Drip-Proof with Extra Varnish Treatments. Same as standard open drip-proof except with extra dips and bakes to increase moisture resistance of the insulation system.

Totally Enclosed. The windings and internal parts are protected by the frame enclosure which prevents free exchange of ambient air. Recommended for dirty or outdoor applications where high reliability and long life are prime considerations (Fig. 16–4).

Totally Enclosed, Severe Duty. Offers same enclosure as above, but with special features—designed for use in chemical atmospheres or extremes of moisture and humidity. Table 16–2 is a guide to the proper selection of the enclosure and/or winding treatment, on applications where chemicals are encountered, and/or where mechanical protection from dust is required.

Each construction is ranked as to its ability to withstand the particular condition. Note that the drip-proof motor with extra varnish treatment is well suited to high humidity.

Fig. 16–4. Totally enclosed, nonventilated, split-phase, resilient mounting base motor. *(Courtesy Westinghouse)*

Mechanical Protection. Where clogging materials are present in severe proportions, the air gap of open motors may become clogged. Therefore, the recommendation is a totally enclosed motor.

Resistance to Chemicals. The first recommendation for any type of atmosphere containing chemicals, acids, bases, solvents, etc., should be severe-duty enclosed construction.

Tropical Protection. Specifications for motors to be used in a hot and humid location may call for tropical protection or tropical insulation. This will be assumed to mean that the windings must be specially

Table 16–2. Rating Enclosures

Enclosure and/or Winding Treatment	Humidity Resistance	Mechanical Protection	Resistance to Chemicals
Open, Drip-Proof	4	5	4
Open, Drip-Proof with Extra Varnish Treatment	2	4	3
Moisture-Sealed	1	3	3
Totally Enclosed	3	2	2
Severe-Duty	2	1	1

Rank: 1 = High, 5 = Low

protected against moisture and fungus and able to operate with normal life expectancy in higher-than-normal ambient temperatures up to 65°C.

Motors specified for tropical use will be supplied with extra dips and bakes of insulating varnish and a 65°C ambient insulation system. Special treatment for fungus proofing is not required, as the materials used in the insulation system are resistant to fungus growth.

Anti-Fungus Treatment. The anti-fungus requirement can be met by either supplying an insulation system which will not support fungus growth or a special anti-fungus varnish. Since either may be used, depending on the rating, do not specify anti-fungus varnish but request anti-fungus treatment.

Guide to Selecting the Right Motor

AC motors can be divided by power supply into two major electrical categories, polyphase and single-phase. The four important selection criteria to consider are horsepower, voltage, speed, and phase.

Over 90 percent of the motors used in the United States are single-phase. The most popular electrical types of single-phase motors for general use are the capacitor-start and the split-phase.

The split-phase motor has two windings, start and run, that are energized to start the motor. The start winding is cut out of the circuit at about 75 percent of operating speed. The capacitor-start motor is almost identical to the split-phase motor but delivers two or three times the starting torque per ampere of current.

Use

The lower-cost split-phase motor is the logical motor to choose for applications where the starting load is light, such as fans and blowers, or where the load is applied after the motor has reached operating speed; for instance, on saws and drill presses. Capacitor-start motors are necessary, of course, on applications such as conveyors where heavy loads must be started.

To make sure there is a correct motor for every job, single-phase motors are available in two types: general-purpose and special-service.

Table 16–3 shows the differences between the four basic single-phase motors.

Motor Speed

Almost all 60-Hz split-phase and capacitor-start motors operate at one of these full-load speeds: 3450 rpm (two-pole), 1725 rpm (four-pole), 1140 rpm (six-pole), or 860 rpm (eight-pole).

Table 16–3. Typical Characteristics of AC Motors

Motor Types	SPLIT-PHASE General-purpose	SPLIT-PHASE Special Service	CAPACITOR-START Special Service	CAPACITOR START General-purpose	POLYPHASE 1 HP & Below
Starting Torque (% Full Load Torque)	130%	175%	250%	350%	275%
Starting Current	Normal	High	Normal	Normal	Normal
Service Factor (% of Rated Load)	135%	100%	100%	135%	135%
Comparative Price Estimate (Based on 100% for Lowest Cost Motor)	110%	100%	135%	150%	150%
Remarks	Low starting torque. High service factor permits continuous loading—up to 35% over nameplate rating. Ideal for applications of medium starting duty.	Moderate starting torque, but has service factor of 1.0. Apply where load will not exceed nameplate rating for any extended duration of time. Because of higher starting current, use where starting is infrequent.	High starting torque but 1.0 service factor. Use only where load will not exceed nameplate rating for any extended duration of time. Starting current is normal.	Very high starting torque. High service factor permits continuous loading up to 35% over nameplate rating. Ideal for powering devices with heavy loads, such as conveyors.	Normal start current for polyphase is low compared to single-phase motors. High starting ability. High service factor permits continuous loading up to 35% over nameplate rating. Direct companion to general-purpose capacitor-start motor.

Permanent-Split-Capacitor Motors (KCP) are normally designed for direct-drive fan and blower applications. Because of the uniqueness of design and applications, these types of motors are excluded from the above table. (Courtesy General Electeric)

Generally speaking, both motor price and physical size increase as rated rpm decreases. This means that savings can be achieved by selecting the highest-speed motor that will drive the device yet still remain within the practical limits of a 5 to 1 or 6 to 1 speed-reduction ratio. Obviously, such savings are most often available in belt-drive applications.

Bearings

All-angle sleeve-bearing motors can be used over a wide range of applications where moderate thrusts are encountered. They can be mounted in any position and they are quieter and more economical than ball-bearing motors. The motor applications which demand ball-bearing motors are in powering devices which create high axial and/or radial thrust (Fig. 16–5).

Installation

Power Supply/Connections

Connect the motor for the desired voltage and rotation according to the connection diagram on the nameplate or in the terminal box (Fig. 17–6).

Voltage, frequency, and phase of the power supply should be consistent with the motor nameplate rating. The motor will operate satisfactorily on voltage within 10 percent of nameplate value or frequency within 5 percent (combined variation not to exceed 10 percent) of nameplate value.

Operation of 230-Volt Motors on 208-Volt Systems

Motors rated 230 volts will operate satisfactorily on 208-volt systems on most applications requiring nominal starting torques. Starting and maximum running torques of a 230-volt motor will be reduced approximately 25 percent when operated on 208-volt systems. Fans, blowers, centrifugal pumps, and similar loads will normally operate satisfactorily at these reduced torques. Where the application torque requirements are high, it is recommended that the next-highest-horsepower 230-volt motor or a 200-volt motor be used. External motor con-

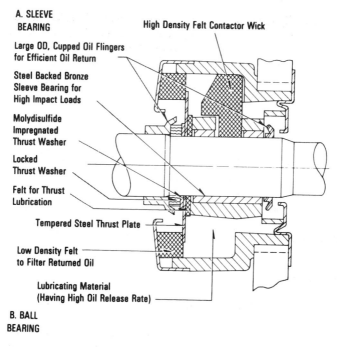

A. SLEEVE
 BEARING

High Density Felt Contactor Wick

Large OD, Cupped Oil Flingers
for Efficient Oil Return

Steel Backed Bronze
Sleeve Bearing for
High Impact Loads

Molydisulfide
Impregnated
Thrust Washer

Locked
Thrust Washer

Felt for Thrust
Lubrication

Tempered Steel Thrust Plate

Low Density Felt
to Filter Returned Oil

Lubricating Material
(Having High Oil Release Rate)

B. BALL
BEARING

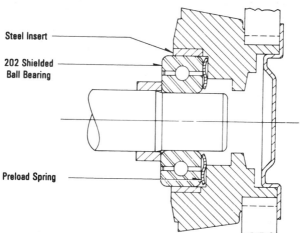

Steel Insert

202 Shielded
Ball Bearing

Preload Spring

Fig. 16–5. (A) shows the sleeve bearing and how it is quieter running than the ball-bearing motor shown in (B).

Fig. 16–6. The direction of rotation of this motor can be changed by slipping off the slip-on connectors and changing them to fit the directions given on the nameplate for the direction of rotation. *(Courtesy Westinghouse)*

trols for 230-volt motors on 208-volt systems should be selected from 230-volt nameplate data.

Safety Precautions

1. Use safe practices when handling, lifting, installing, operating, and maintaining motors and motor-operated equipment.
2. Install motors and electrical equipment in accordance with the *National Electrical Code (NEC)* or sound local electrical and safety codes and practices, and, when applicable, the Occupational Safety and Health Act (OSHA).
3. Ground motors securely. Make sure that grounding wires and devices are, in fact, properly grounded. *Caution: Failure to ground a motor properly may cause serious injury to personnel.*

> ### Caution
>
> Motors subjected to overload, locked rotor, current surge, or inadequate ventilation conditions may experience rapid heat buildup, presenting risks of motor damage or fire. To minimize such risks, use of motors with proper overload protectors is advisable for most applications.
>
> Do not use motors with automatic-reset protectors where automatic restarting might be hazardous to personnel or to equipment. Use motors with manual-reset protectors where such hazards exist. Such applications include conveyors, compressors, tools, and most farm equipment.

4. Before servicing or working on or near motors or motor-driven equipment, disconnect the power source from the motor and accessories (Fig. 16–7).
5. Remove shaft key from keyways of uninstalled motors before energizing the motor. Be sure the keys, pulleys, fans, and other attached parts are fully secured or installed on the motors before energizing the motor.
6. Make sure fans, pulleys, belts, and other parts are properly

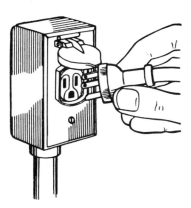

Fig. 16–7. Remove the plug from the receptacle before working on a motor. If it is connected to a switchbox, make sure the switch is off. Note there is a third hole here which represents the grounding wire that connects to the outside frame of the motor. *(Courtesy Sears)*

guarded if they are in a location that could be hazardous to personnel.

7. Provide proper safeguards against failure of motor-mounted brakes, particularly on applications involving overhauling loads.

8. Provide proper safeguards on applications where a motor is mounted on or through a gear reducer to a holding or over-hauling application. Do not depend on gear friction to hold the load.

9. Do not lift the motor-driven equipment with motor-lifting means. If eyebolts are used for lifting motors, they must be securely tightened and the direction of the lift must not exceed a 15° angle with the shank of the eyebolt.

Location and Motor Enclosures

Open, drip-proof motors are designed for use in areas that are reasonably dry, clean, well ventilated, and usually indoors. If installed outdoors, it is recommended that the motor be protected with a cover that does not restrict flow of air to the motor.

Totally enclosed motors are suitable for use where exposed to dirt, moisture, and most outdoor locations, but not for very moist or for hazardous (atmosphere filled with explosive vapor dust) locations.

Severe-duty enclosed motors are suitable for use in corrosive or excessively moist locations.

Explosion-proof motors are made to meet Underwriters Laboratories Standards for use in hazardous (explosive) locations shown by the UL label on the motor (Fig. 16–8).

Certain locations are hazardous because the atmosphere does or may contain gas, vapor, or dust in explosive quantities. The *National Electrical Code* (*NEC*) divides these locations into classes and groups according to the type of explosive agent which may be present. Listed below are some of the agents in each classification. For complete listing, see Article 500 of the *National Electrical Code*. A newly revised code is published every three years.

Class I (Gases, Vapors)

Group A — Acetylene (*Note:* Motors are not available for this group).

Group B — Butadiene, ethylene oxide, hydrogen, propylene oxide (*Note:* Motors are not available for this group).

Fig. 16–8. **Explosion-proof motor** *(Courtesy Westinghouse)*

Group C — Acetaldehyde, cyclopropane, diethyl ether, ethylene, isoprene.

Group D — Acetone, acrylonitrile, ammonia, benzene, butane, ethylene dichloride, gasoline, hexane, methane, methanol, naphtha, propane, propylene styrene, toluene, vinyl acetate, vinyl chloride, xylene.

Class II (Combustible Dusts)

Group E — Aluminum, magnesium, and other metal dusts with similar characteristics.

Group F — Carbon black, coke, or coal dust.

Group G — Flour, starch, or grain dust.

Ambient temperature around motors should not exceed 40°C unless the motor nameplate specifically permits a higher value.

Maintenance

Motors properly selected and installed are capable of operating for many years with a reasonably small amount of maintenance.

Before servicing motors and motor-operated equipment, disconnect the power supply from the motors and accessories. Use safe working practices during servicing of equipment.

Clean motor surfaces and ventilation openings periodically, preferably with a vacuum cleaner. Heavy accumulations of dust and lint will result in overheating and premature failure of motors.

Lubrication. Motors are lubricated at the factory to operate for long periods under normal service conditions without relubrication. Excessive or too frequent lubrication may damage the motor. Follow instructions furnished with the motor—usually on the nameplate or terminal box cover or on a separate instruction.

If instructions are not available, relubricate as follows: For sleeve-bearing motors, add electric motor oil (or SAE #10 or #20 nondetergent oil) after three years of normal or one year of heavy-duty service. For light-duty applications, add oil after 25,000 hours of operation. Ball-bearing motors of the newer types have sufficient lubrication for many years of operation. The time period between relubrication can

Fig. 16–9. To dry wet motor, disconnect and place in oven set at lowest temperature. Open door and leave in oven until steaming stops.

vary from 10 years to 9 months, depending on the ambient temperature and the type of service. For specific regreasing recommendations, refer to the instruction provided with the motor or contact the manufacturer's service center locally.

Wet Motors. If a motor gets wet, disconnect and place in an oven set at its lowest temperature and leave it until it stops steaming (Fig. 16–9).

PART SIX

Electronically-Operated Motors

CHAPTER 17

Steppers and Synchronous Motors

With the advent of the computer and programmable controllers, there is a need for electric motors that can be controlled by using electronics. These motors have taken various forms, but still rely upon the basic principles of motor operation with some additional characteristics developed for electronic signal control purposes.

Robotics has demanded more accuracy in motor and motion control. Programmable controllers have also provided the necessary signals for making small precision motors do what is intended as an end product or motion.

Some of these motors are very expensive and are found in precision measuring devices and strip recorders. They are usually made to require little or no maintenance; however, in some instances it may become necessary for you to open the motor and check for various conditions. In order to know what you are doing, it is also best to know the circuitry, the intended purpose of the motor, and how it is supposed to operate to determine if it is malfunctioning.

The intent of this chapter is to introduce some of the characteristics and applications of specialty motors that use electronics to control their actions.

Shaded-Pole Unidirectional Synchronous Motors

The shaded-pole, one direction of rotation (unidirectional) synchronous motor is used for timing purposes (Fig. 17–1), such as driving clocks, elapsed time indicators, repeat cycle timers, stop clocks, and potentiometers. This type of motor is also found as a drive device for encoder disc chart recorders and medical instrumentation, and as an agitator drive for blood analyzers. It is an inexpensive type of motor that can be utilized in visual displays as a driver for cycle changes. It is also used for fiber optic displays, an oil pump drive on copying machines and as a code wheel drive on remote fire alarm systems.

The instant start-stop characteristic of this motor is useful for integrating time intervals. It provides an ideal drive for elapsed time indicators used for billing on a time-usage basis, for stop clocks measuring precise time intervals and for chart drives. In many applications, the fast start-stop ability eliminates the need for a clutch. When voltage is applied, the motor accelerates from dead stop to full speed in less than 0.01 of a second with no load. For this reason, the inertia of any member rigidly coupled to the output shaft should be kept to a minimum or a resilient coupling used.

Direction of rotation may be left (ccw) or right (cw) and it may be obtained with a 115-volt or 20-volt coil. Note the two leads for this

Fig. 17–1. Shaded-pole unidirectional synchronous motor. *(Courtesy of Hayden Switch & Instrument, Inc.)*

particular type of motor. The torque available for various voltages and speeds is shown in Table 17–1.

Capacitor Start and Run Bidirectional Synchronous Motor

This motor looks like the previous one except that it has three wires coming from the case. This motor is easily reversed by switching a capacitor from one coil to the other. The small size of the motor enables one to design equipment to fit the motor and capacitor (Fig. 17–2). The 115-volt, 60-Hz motor requires a 0.35-MF capacitor while the 20-volt, 60-Hz type uses a 9.0-MF capacitor.

Fast start-stop and reversing characteristics eliminate the need for clutching, braking, or pre-starting in many applications. This motor is ideally suited for control functions where shafts, potentiometers, indicators and the like must be precisely positioned. The motor can "inch" in either direction in small increments and stop without coasting.

This type of motor can be found in chart drives with portable pen drives for plotting curves. Spectrographs use them to drive potentiometers for speed control analyzers. Remote temperature control on master panels for heat treating also use this type of motor. Industrial processing equipment will use it where remote speed control is called for. Chart recorders use it to drive the code wheel. These motors are also used for ink flow control on printing presses and lens system drives for

Table 17–1. Speed-Torque, Shaded Pole Unidirectional Synchronous Motor

Volts	Output Speed (RPM)				Rotation
	360	300	60	3.6	
115	0.08 Oz-In	0.1 Oz-In	0.5 Oz-In	5.0 Oz-In	Left or Right (ccw) (cw) Depends on Motor Selected
20	0.08 Oz-In	0.1 Oz-In	0.5 Oz-In	5.0 Oz-In	Left or Right Depends on Motor Selected

Oz-In = Ounce-Inch (unit for torque measurement)
Gear train for this type of motor is designed for no more than 5.0 oz-in. Do Not Exceed.
Ambient Temperature 0° to +50°C continuous duty.

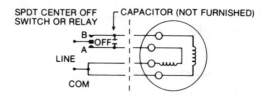

Fig. 17–2. Capacitor start and run bidirectional synchronous motor wiring diagram. *(Courtesy Hayden Switch & Instrument, Inc.)*

focus control on 35mm slide and 35mm motion picture projectors. Industrial packing equipment uses the motor for pot control.

The typical drive circuit is shown in Fig. 17–3. Note the triacs and the two inputs A and B. In Table 17–2, the A and B represent these two inputs. Power input for these motors is about two watts. Acceleration from dead stop to full speed is 0.01 second with no load.

Single-Phase Stepper Motor (Unipolar Drive)

This motor has only two leads (Fig. 17–4). It operates in only one direction depending on the motor selected, either left rotation or right rotation, but it is not reversible as made (Fig. 17–5). A motor of this type comes in handy for a number of uses as it is small and converts electrical pulses into discrete angular steps of the output shaft without

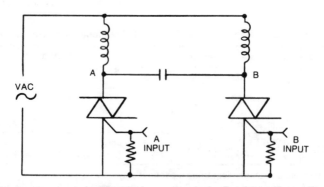

Fig. 17–3. Typical drive circuit for capacitor start and run bidirectional synchronous motor. *(Courtesy Hayden Switch & Instrument, Inc.)*

Table 17–2. Step-Angle, Torque and Pulse Rate

Step Angle	Torque (Oz-In)	
	1–20 Hz	40 Hz
36°	0.3	0.2
30°	0.3	0.2
6°	1.7	1.0
3.6°	3.0	2.0

Motors are available on special order to operate on 12 volts.

need for control logic (Fig. 17–6A). For each impulse, the rotor turns 360°; 180° when power is applied (external pulse). By means of magnetic detenting, the rotor turns an additional 180° when power is then removed (internal pulse). No power is consumed between pulses, making the motor ideal for use in battery-operated systems.

This motor has a special capability for operation in applications where the power supply is limited and command pulses are widely spaced, since a low duty cycle will reduce the average power drain to a few milliwatts. For example, the motor could drive a battery-operated digital clock for automotive or aircraft applications where average power consumption must be low. A counter reading 23 hours 59 minutes and seconds in increments of 5 seconds could be driven by the two-wire stepper motor geared to step the seconds drum in 12°, or 5

Fig. 17–4. Single-phase stepper motor, unipolar drive. *(Courtesy Hayden Switch & Instrument, Inc.)*

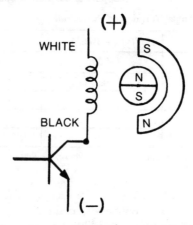

Fig. 17–5. Typical drive circuit for single-phase stepper. *(Courtesy Hayden Switch & Instrument, Inc.)*

second increments, from an accurate pulse source such as a crystal-controlled oscillator circuit.

The peak power requirement would be approximately 2 watts for a pulse duration of 0.010 seconds or an average power consumption of only 4 milliwatts.

$$\frac{2 \times 0.010}{5.0} = 4\text{mW}$$

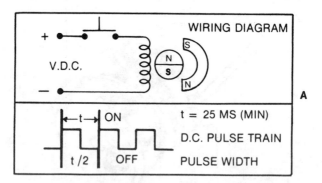

Fig. 17–6. (A) Wiring diagram and pulse needed for operation. (B) Placement of capacitor in circuit. *(Courtesy Hayden Switch & Instrument, Inc.)*

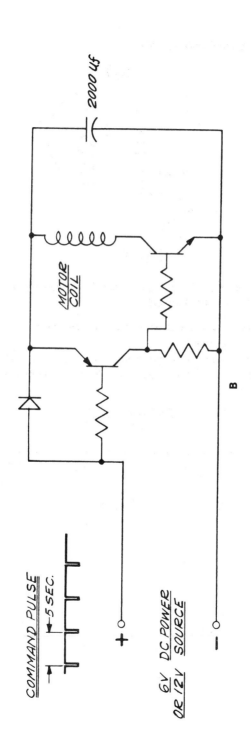

In cases where the power supply is not capable of providing 2 watts peak power but is capable of handling the low 4mW average wattage requirement, a system developed by Tri-tech, Inc. might be of interest. In the circuit shown in Fig. 17–6B, a capacitor is charged at a low rate between pulses and discharged into the stepper motor by a short negative-going command pulse from a time base of 5-second intervals. In the interval between pulses, the capacitor is recharged, storing up energy to drive the stepper motor when the next signal pulse is received.

The two-wire stepper motor is capable of many modes of operation. The significant thing to remember is that energy is consumed only while the stepper motor coil is energized and only for a short period of time as low as 10 milliseconds. The average power consumed depends on the pulse interval. If this is long with respect to the 10 ms interval, then the average power will be quite low. This type of motor can find use in disc or strip chart recorders, fire alarm reporting systems and other devices operated from a battery power source.

The step rate is from 0 to 80 total pulses per second (pps), 40 external and 40 internal, with a 12.5-millisecond minimum pulse width for each. For maximum motor efficiency, a symmetrical input (equal ON and OFF times) is always desirable (Fig. 17–7). Internal pulse is produced by the collapsing magnetic field generated by the first (external) pulse.

Uses for this type of motor include electronic speedometer/odometer drive, bidirectional tape footage counter using two motors with mechanical differential, driver for watt-hour meters, counter drive for flow meters, rotary stepping switch drives, marine odometers and trip odometers, as well as recorder tape drives. They operate well on 24 volts DC or 12 volts DC depending on the coil used in the motor.

Single-Phase Stepper Motor (Bipolar Drive)

The unidirectional-series, single-phase, bipolar-drive stepper motor exhibits all the qualities of the synchronous motor and even looks like it. Its low wattage, high frequency response, compact size, and low cost are significant when considering it for various purposes. Pulses may be derived from simple DPDT circuitry (Fig. 17–8) arranged to reverse the polarity of the voltage applied to the two motor

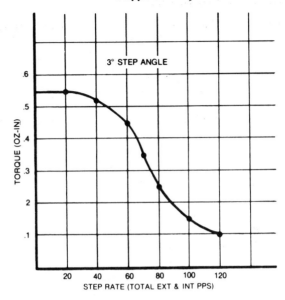

Fig. 17–7. Step rate vs. torque graph. *(Courtesy Hayden Switch & Instrument, Inc.)*

leads (Fig. 17–9). This motor is available in both 12- and 24-volt sizes. They will rotate either left or right depending upon the design. Fig. 17–10 shows the 3° step angle on the torque-step rate chart.

There is the 18°, 15°, 3°, and 0.18° step-angle per pulse with a

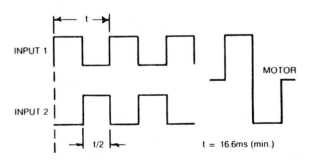

Fig. 17–8. Typical pulse for single-phase stepper motor, bipolar drive. *(Courtesy Hayden Switch & Instrument, Inc.)*

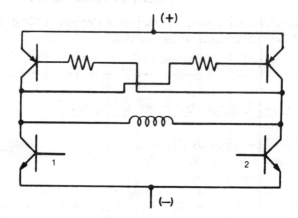

Fig. 17–9. Typical drive circuit for single-phase stepper motor, bipolar drive. *(Courtesy Hayden Switch & Instrument, Inc.)*

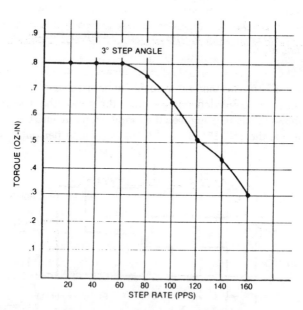

Fig. 17–10. Graph for torque vs. step rate for single-phase stepper motor, bipolar drive. *(Courtesy Hayden Switch & Instrument, Inc.)*

maximum torque of 5 oz-in. The maximum step rate is 120 pulses per second. Input power is about two watts.

This type of motor is used for paper tape drive for thermal printers, chart drives, and DC counter drives.

Two-Phase Stepper Motor (Unipolar Drive)

This motor resembles the previous one except it has three leads (white-black-black) coming from the enclosure, and it provides an economical solution to instrumentation problems requiring a small compact drive system. For each pulse, the rotor revolves 180° when coil B is energized, and another 180° when coil A is energized in sequence. The step rate is from 0 to 120 pps.

Simple control circuitry for this type of motor is shown in Fig. 17–11. Note where the black and white leads are connected to the circuitry, with the black leads each connected to a collector of a different transistor.

The wiring diagram indicates the coil and control switch arrangement in reference to the DC input source. The typical pulse is shown in Fig. 17–12.

This type of motor is made for 12 or 24 volts depending upon the selection of coils. Maximum torque is 5 oz-in. Step angle per pulse is 18°, 15°, 3°, or 0.18° depending on the motor selected. Left or right rotation is available.

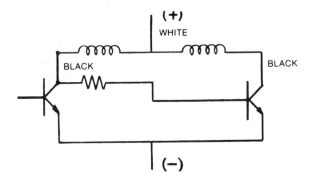

Fig. 17–11. Typical drive circuit for two-phase stepper motor, unipolar drive. *(Courtesy Hayden Switch & Instrument, Inc.)*

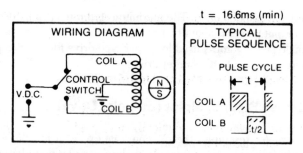

Fig. 17–12. Wiring diagram and typical pulse sequence for two-phase stepper. *(Courtesy Hayden Switch & Instrument, Inc.)*

Fig. 17–13 shows the torque for a 3° step angle motor with variations in torque as the step rate is changed from 20 to 160 pps.

Some of the practical applications for the motor are as counter drivers, indexing control requirements, drivers for rotary stepping

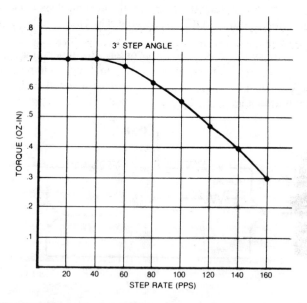

Fig. 17–13. Graph for torque vs. step rate for two-phase stepper motor. *(Courtesy Hayden Switch & Instrument, Inc.)*

switches, marine odometers, seismograph recorders, and as a driver for watt-hour meters.

Four-Phase Stepper Motor (Unipolar Drive)

This stepper motor has two center tapped coils and two permanent magnet rotors in a tandem magnetic structure. Each half-coil is considered a separate phase (Fig. 17–14).

The motor resembles all the previous ones in appearance, but has more wires coming from the enclosure. The wires are white for the common, green, black, blue, and red (Fig. 17–15). Energizing two phases at a time in the sequence shown in Fig. 17–15 will cause the rotor to turn in 90° steps. This may be done conveniently with two flip-flop circuits arranged to provide the desired sequence of coil excitation. With the appropriate logic circuitry, the reversible stepper can be used for step servo positioning, for instrument drives in self-balancing potentiometer circuits, for constant or variable speed drives for tapes, charts, cams, actuators, or drums and for similar applications. Fig. 17–16 shows the pulse sequence for clockwise and counter-clockwise rotation. Power input is about 1.5 watts per phase.

The motor is available in either 12 or 24 volts. The step angle per pulse is 9°, 7.5°, 1.5° or 0.09° depending on the motor selected. Note in Fig. 17–17 that the maximum torque is about 1.4 oz-in for the 1.5° series of motors. Maximum step rate for this motor is 240 pps.

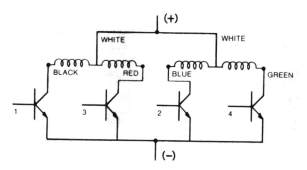

Fig. 17–14. Typical drive circuit for four-phase stepper motor, unipolar drive. *(Courtesy Hayden Switch & Instrument, Inc.)*

1
WHITE
3
RED
BLACK
2
BLUE
N
S
WHITE
GREEN
4
COM

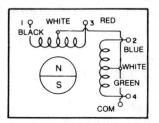

Fig. 17–15. Wiring code for four-phase stepper motor, unipolar drive. *(Courtesy Hayden Switch & Instrument, Inc.)*

You can understand the popularity of these motors when you consider that they weigh 2 to 2.5 ounces with a diameter of only 1 inch and a maximum width of about 1.25 inches.

Open-Frame Synchronous and Stepper Motor

When more torque is needed, the open-frame motor is used, as it is larger and capable of driving heavier loads (Figs. 17–18 and 17–19). All types of these motors have a barium ferrite rotor and a laminated silicon steel stator structure forming two split poles. The motor shaft and the output shaft are journaled in sintered bronze bearings. The gears are molded nylon, turning on fixed studs driven into the gear train plates. The motor pinion and the final gear have brass inserts and are driven on the motor shaft and the output shaft, respectively. This provides a strong gear train capable of carrying 10 oz-in continuous duty or 30 oz-in intermittent duty.

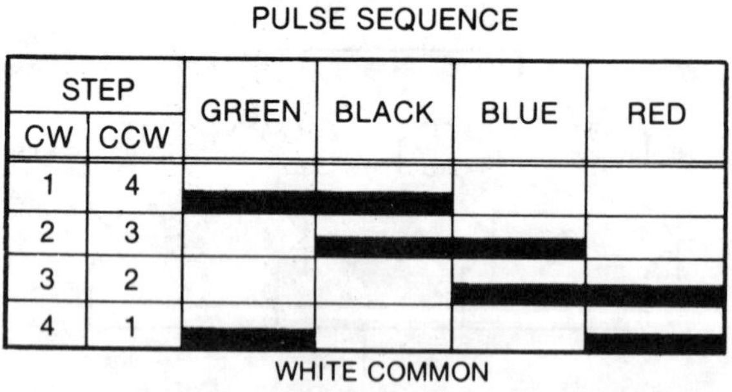

PULSE SEQUENCE
STEP
CW
CCW
GREEN
BLACK
BLUE
RED
1 4
2 3
3 2
4 1
WHITE COMMON

Fig. 17–16. Pulse sequence for four-phase stepper motor, unipolar drive. *(Courtesy Hayden Switch & Instrument, Inc.)*

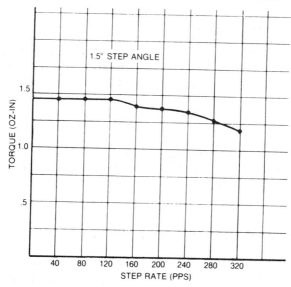

Fig. 17–17. **Graph for torque vs. step rate for four-phase stepper motor, unipolar drive.** *(Courtesy Hayden Switch & Instrument, Inc.)*

Basically, the open-frame 2-wire stepper motor (Fig. 17–18) is the same as the series 2-wire motor previously mentioned (Fig. 17–4) and enclosed in a 1-inch diameter can. The 360° rotation per pulse cycle is the same, as is the 2-wire control circuitry (Fig. 17–20). The applications are also the same, except this motor is used where the added

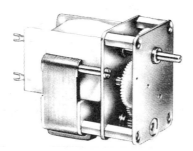

Fig. 17–18. **Open-frame unidirectional shaded-pole stepper motor.** *(Courtesy Hayden Switch & Instrument, Inc.)*

Fig. 17–19. Open-frame reversible shaded-pole step servo motor. *(Courtesy Hayden Switch & Instrument, Inc.)*

weight and space needed for mounting is not a problem and where more torque is needed to operate the driven device.

Keep in mind that the rotor and output shaft bearings are sintered bronze and vacuum impregnated so no further lubrication is needed. The integral gear reducer, with nylon gears and pinions, provides ex-

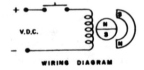

WIRING DIAGRAM

TYPICAL PULSE SEQUENCE

Fig. 17–20. Wiring diagram and typical pulse sequence for a 2-wire stepper motor. *(Courtesy Hayden Smith & Instrument, Inc.)*

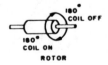

ROTOR

tremely long life and requires no lubrication. The lubricant supplied is usually for operation over an ambient temperature range of 0–71°C. Special lubrication can be obtained for operation over a wider ambient temperature range.

Stepper motors, as well as relays, solenoids and other devices with coils, are inductive loads. That means arc suppression should be provided for contact or transistor protection and to minimize the generation of transients. Simple diode suppression across the coil should not be used as this adversely affects performance. Instead, a combination of a blocking diode and a zener diode in series across the coil is recommended. A typical circuit for the stepper and voltage and current waveforms for one pulse cycle are shown in Fig. 17–21.

The average weight for these motors is 6 ounces, and the power consumption averages 3 watts on 24 volts DC. Fig. 17–22 shows the dimensional data for this type of open-frame motor. Table 17–2 (p. 345) illustrates the torque in oz-in for the various step angle motors. Note how the 1 to 20 Hz and 40 Hz pulsing makes a difference in the torque provided. These motors are available in right rotation or left rotation, but are not reversible.

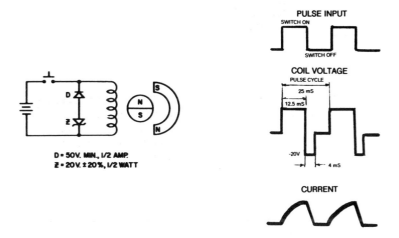

Fig. 17–21. Arc suppression and pulses for a stepper motor. *(Courtesy Hayden Switch & Instrument, Inc.)*

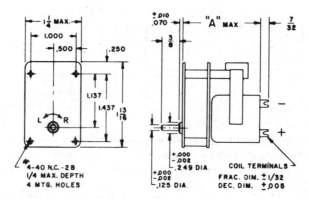

Fig. 17–22. Dimensional data for a 2-wire stepper motor. *(Courtesy Hayden Switch & Instrument, Inc.)*

Reversible Step Servo Motor

The series step servo motor has two permanent magnet rotors in a tandem laminated magnetic structure with two center tapped coils (Fig. 17–19). For purposes of identification, each half coil is considered a separate coil (Fig. 17–23). With connections shown in Fig. 17–24, energizing the coils individually in sequence 1-2-3-4 will cause the rotor to turn right in 90° steps for one revolution (360°).

Normally, this type of operation is not practical since it is simpler to drive the step servo with two flipflop circuits that at any instant energize two coils. This mode of operation is shown in Fig. 17–25. The sequence of operation then becomes 1-4, 1-2, 2-3, 3-4 for right rotation and 1-4, 3-4, 2-3, 1-2 for left rotation.

The step servo has no magnetic detent when deenergized and consequently the rotor position does not shift when power is removed from the system. The unit will operate with a rated load up to 50 com-

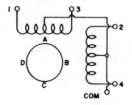

Fig. 17–23. Wiring diagram for reversible step servo motor. *(Courtesy Hayden Switch & Instrument, Inc.)*

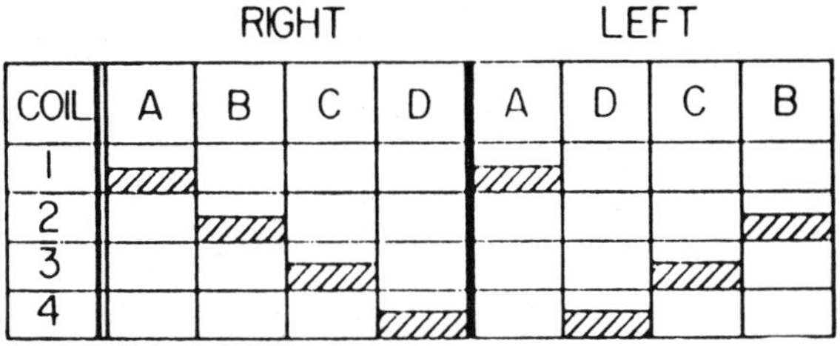

COIL	RIGHT				LEFT			
	A	B	C	D	A	D	C	B
1	/////				/////			
2		/////						/////
3			/////				/////	
4				/////		/////		

Fig. 17–24. Mode of operation, sequence of operation. *(Courtesy Hayden Switch & Instrument, Inc.)*

plete cycles per second or 200 changes of state per second each resulting in a 90° rotor step. Maximum slewing rate is approximately twice this figure.

The motors are available in 9°, 7.5° and 1.5° step angles. Typical torque output with the 9° step angle motor varies from 0.1 oz-in at 200 steps per second to 0.2 oz-in at 1 step per second (Table 17–3). Power required is 8 watts nominal.

With appropriate solid state logic circuitry, the series step servo motor provides a reliable variable-speed drive for strip chart recorders and other devices requiring a range of speed on the order of 200:1 or higher. If an accurate frequency standard is available, solid state divider circuits will enable stepper operation at predetermined submultiples of the referenced frequency.

Again, with the appropriate solid state logic circuitry, the series stepper can be operated as a reversible step servo motor for instrument

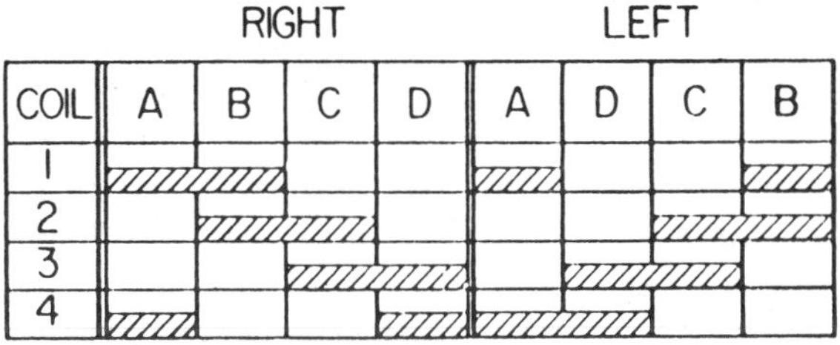

COIL	RIGHT				LEFT			
	A	B	C	D	A	D	C	B
1	/////////				/////			/////
2		/////////						/////////
3			/////////			/////////		
4	/////			/////	/////////			

Fig. 17–25. Mode of operation, sequence of operation. *(Courtesy Hayden Switch & Instrument, Inc.)*

Table 17–3. Step Angle/Dimension B

Voltage 24 Volts DC ± 19%			
Step Angle per change of state	9°	7.5°	1.5°
Dimension B (Fig. 17–26)	0.672″	0.672″	1.137″

drives in self-balancing potentiometer circuits and similar applications.

Fig. 17–26 shows the coil connections and the dimensions of the motor. Left rotation, of course, refers to counter-clockwise (CCW) rotation and the right rotation refers to clockwise rotation (CW) when the motor shaft is viewed from the end as shown in Fig. 17–26.

High-Torque Reversible AC Motor

This is an instantly reversible motor. Fig. 17–19 shows what it looks like, and Fig. 17–27 shows the wiring diagram. It is reversed by switching the capacitor from one coil to the other. This makes it ideal for positioning potentiometers, shafts, indicators, and similar service

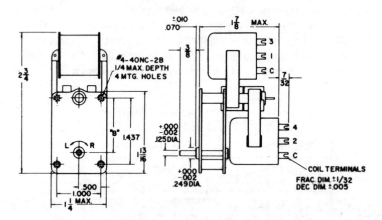

Fig. 17–26. Dimensional data for reversible step servo motor. *(Courtesy Hayden Switch & Instrument, Inc.)*

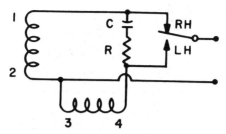

Fig. 17–27. Wiring diagram for high-torque reversible AC motor. *(Courtesy Hayden Switch & Instrument, Inc.)*

since it starts and stops within one cycle of AC power. The series motor can "inch" in either direction in small increments and stop without coast. With split coils, this motor becomes a step servo motor capable of bidirectional operation up to 200 pps.

This motor operates on 115 volts ± 10% and consumes 2.5 watts. Torque at 3600 rpm is 0.07 oz-in. The rotor and output shaft bearings are porous bronze and are lubricated for life at the factory for operation over an ambient temperature range of 0° to 71°C. The capacitor is 0.27 MF and the resistor used in the circuit is 4.5k ohms at 5 watts. When checking the coil terminal numbers, start at the top with 1 and 2. Coils 3 and 4 are on the bottom. Table 17–4 shows the speed and torque for four different speed motors. Gear train is rated at 10 oz-in continuous and 30 oz-in intermittent duty.

Table 17–4. Speed/Torque Characteristic for 115V 60 Hz Motors

Speed	Torque
60 rpm	4 oz-in
30 rpm	8 oz-in
20 rpm	12 oz-in
10 rpm	24 oz-in intermittent

Gear trains are rated at 10 oz-in continuous and 30 oz-in for intermittent duty.

3-Wire Stepper Motor (Open Frame)

The 3-wire stepper is slightly different from the 2-wire. Take a look at the wiring diagram in Fig. 17–28. Magnetic detenting occurs at 0° and 180° rotor positions whether or not the windings are energized.

The 3-wire stepper produces 360° rotation with each pulse. The typical pulse sequence is also shown in Fig. 17–28. This comes in handy when used in pulse counters, stepping switches, pulse storage devices and remote positioning. Note how arc suppression is accomplished in Fig. 17–29. Simple diode suppression across the coils should not be used as this adversely affects performance. Instead, a combination of blocking diodes and a zener diode is recommended. Note the typical voltage and current waveforms for one pulse cycle.

Power consumed is 4 watts at 60 Hz and double that or 8 watts at 1 Hz. The motor weighs 6 ounces and can be obtained for 12-volt operation when specially ordered.

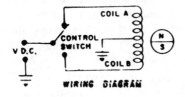

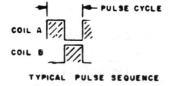

Fig. 17–28. Wiring diagram and pulse sequence for a 3-wire stepper motor. *(Courtesy Hayden Switch & Instrument, Inc.)*

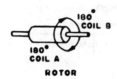

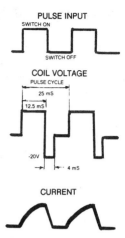

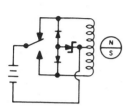

Fig. 17–29. Arc suppression and pulses for operation of 3-wire stepper motor. *(Courtesy Hayden Switch & Instrument, Inc.)*

Table 17–5 shows the four step angles available in these motors and the torque generated at two different frequencies.

Fig. 17–30 shows that the center tap is the bottom coil terminal and not the middle one. Dimension A is either 1.4375 or 1.750 depending on the step angle. The larger step angles of 36° and 30° have the smaller size while the 6° and 3.6° step angle motors are a little larger at 1.75 inches. Dimension B is 0.672 inch on the larger step angle motors and 1.137 inches on the 6° and 3.6° motors.

Table 17–5. Step Angle/Torque

Step Angle	Torque (Oz-In)	
	1–20 Hz	60 Hz
36°	0.6	0.3
30°	0.6	0.3
6°	3.5	1.5
3.6°	6.0	3.0

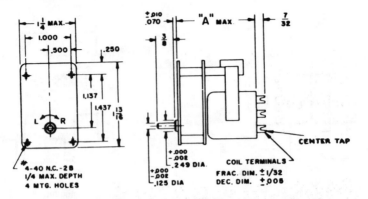

Fig. 17–30. Dimensional data for 3-wire stepper motor. Note the center tap location. *(Courtesy Hayden Switch & Instrument, Inc.)*

CHAPTER 18

Motion Control

Terminology is important when working with or working on a motor or any other device. Disc Magnet Stepper Terminology is presented here in an effort to allow persons working with this type of motor a common ground on which to base their discussion be it for the purpose of repair, selection or operation of the motor.

Disc Magnet Stepper Terminology

General Description

It should be remembered that a stepper motor is fundamentally a single or multi-phase synchronous motor, designed for step-by-step operation. Driving such a motor with pulses will result in a more or less uneven rotation, depending on the pulse rate and the load characteristics. However, the drive signal may also be continuous producing a rotating field, the motor then works as a normal synchronous motor. The difference from large synchronous motors is that the excitation field is created by a permanent magnet (disc magnet motor).

Two-Phase Motor — Motor with two phases, electrically shifted by 90°. Each phase may comprise one or more windings, which may be interconnected externally.

Step (or Step Angle)—Angle of rotation for one increment in the elementary step sequence (see 4-step sequence). By definition, one step in a two-phase stepper corresponds to 90 electrical degrees.

Driving Sequence—Sequence of the increments applied to the motor phases in order to establish rotation.

4-Step Sequence—Elementary sequence for two-phase steppers. One cycle corresponds to four full steps and 360 electrical degrees.

A 4-step sequence can be achieved with either one or two phases energized.

Half-Step—Angle of rotation for one increment of an 8-step sequence (two-phase stepper).

8-Step Sequence—Sequence of the drive signals energizing either one or both phases, to get 8 stable positions in a cycle of 360 electrical degrees.

Microstepping—Dividing a full step into a number of intermediate stable positions achieved by using drive voltages or currents of variable levels.

Power Rate

The power rate is a basic figure used mainly in robotics applications. It shows the speed of increase in mechanical power. The power rate is directly related to the time necessary to move a load from one position to another.

Unipolar Drive—The drive voltage is applied between one end and the common point of a center tapped winding. Two switches per phase are needed, and only 50% of the coils are used at any time. The maximum performances of the stepper can not be obtained.

Bipolar Drive—Four switches per phase are needed. Use of 100% of the copper volume results in better performance than in unipolar mode.

unipolar

bipolar

Holding Torque—The maximum steady torque that can be applied to the shaft of an energized motor without causing continuous rotation.

Detent Torque—The maximum torque that can be applied to the shaft of a non-energized motor without causing continuous rotation.

Pull-In Torque—The maximum torque that can be applied to a motor shaft when starting at the pull-in rate.

Pull-Out Torque—The maximum torque that can be applied when running at the pull-out rate.

Pull-In Frequency—The maximum switching rate at which a motor can start or stop without losing steps. This rate is load-dependent.

Pull-Out Frequency—The maximum switching rate at which a motor can operate without losing steps. This rate is load-dependent.

Detent Torque and Rest Position—The remaining torque without excitation is due to the permanent magnet; the friction torque is to be added. The rest positions without current are the same as those with one phase energized.

Dynamic Measurements—Measuring dynamic torque is difficult, since the measuring method itself may influence the system performance. Therefore the accuracy of such measurements is difficult to guarantee, and it is not always possible to compare measurements made with a composite load (friction plus torque opposed to motor torque). For escap®, dynamic measurements are made using pure friction load.

System Performance—A stepper motor is as good as the driver allows it to be. A good motor can give its best performance only with an appropriate driver. By the way, it is always possible to economize on the driving circuit, if the maximum performances of the motor are not needed.

Dual Voltage Control (Bi-level drive)

A dual voltage power driver can be utilized in order to reduce power dissipation without the loss of motor performance. Essentially this scheme uses a high voltage output to initially energize a motor phase. This insures maximum current rise in minimum time. When the current reaches its predetermined maximum value the high voltage is reduced to a lower "maintenance" voltage for the remainder of the step.

Chopper Drive

The chopper drive system limits the motor current by means of voltage modulation. It is a highly efficient means of motor control which permits an initial high voltage (10 to 20 times rated voltage) to energize the motor until the desired current is obtained. The voltage is then switched off until the current decays to a predetermined level after which the voltage is again turned on. This cycle of on-off voltage (chopping) is continued through the drive pulse time. This method of motor drive is well suited for stop start operation.

Disc Magnet Steppers

The elementary sequence of a two-phase stepper has four states corresponding to four steps. Half-steps can be obtained with an 8-step sequence. The possible sequences are shown in Table 18–1.

Full-step and half-step switching sequences control voltage or current signals of constant amplitude, while mini- or micro-step driving sequences imply variable current levels in both phases.

REMINDER: The number of phases, two in a two-phase stepper, should not be confused with the number of coils which is very often four. In this respect, permanent magnet and hybrid steppers differ from variable reluctance steppers, which must have three- or four-phase.

Using Steppers at Their Upper Limits

In continuous operation, coil heating happens to be the primary limiting parameter. The winding temperature must not exceed 155°C (310°F). Temporarily, the phase current may be increased without any risk of demagnetization.

The chopper drive mode—constant current level—leads to additional iron losses which increase with the chopping frequency. This higher temperature risk should not be omitted in thermal calculations. Drive circuit configurations are shown in Figs. 18–1 to 18–3.

Disc Magnet Two-Phase Stepper Motor

This is a motor that weighs only 40 grams (Fig. 18–4). The rotor consists of a thin axially-magnetized disc. The stator design is a short,

Table 18–1. Driving Sequences

2 phases energized

Step No.	Phase A	Phase B	Direction of rotation
1	+	+	C.W. → / ← C.C.W.
2	−	+	
3	−	−	
4	+	−	

1 phase energized

Step No.	Phase A	Phase B	Direction of rotation
1	+	0	C.W. → / ← C.C.W.
2	0	+	
3	−	0	
4	0	−	

Half-step No.	Phase A	Phase B	Direction of rotation
1	+	0	C.W. → / ← C.C.W.
2	+	+	
3	0	+	
4	−	+	
5	−	0	
6	−	−	
7	0	−	
8	+	−	

Drive Circuit Configuration

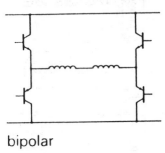

bipolar

Fig. 18–1. Disc magnet stepper bipolar drive circuit. *(Courtesy Stock Drive Products, Div. of Designatronics, Inc.)*

low-loss magnetic circuit resulting in the conservation of the motor characteristics at high speeds. The microstepping motor is designed for applications for high efficiency, a wide speed range, and high power-to-volume and high power-to-weight ratios.

With a single step, the motor can run at 1000 rpm, and features a maximum pullout rate of 10,000 rpm. Theoretically, it can accelerate under maximum current to 465,000 rad/sec^2. This dynamic speed and acceleration allows it to be utilized with a more favorable coupling ratio, and thus eliminate low speed jitters at the load.

Fig. 18–5 is a stepper that has been designed for micro-stepping and therefore has a sinusoidal torque curve and reduced detent torque among its features. It weighs 250 grams with a peak holding torque of 35.7 oz-in.

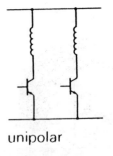

unipolar

Fig. 18–2. Disc magnet stepper unipolar drive circuit. *(Courtesy Stock Drive Products, Div. of Designatronics, Inc.)*

Motor Leads Color Code

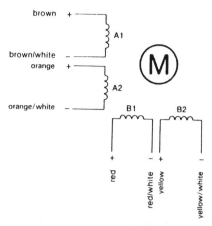

Fig. 18–3. Disc magnet stepper motor leads color code. *(Courtesy Stock Drive Products, Div. of Designatronics, Inc.)*

A larger motor, 600 grams in weight, is shown in Fig. 18–6. It has a peak holding torque of 139.1 oz-in.

PM Two- and Four-Phase Stepper Motors

A permanent magnet rotor-type step motor is shown in Fig. 18–7. The stator is pressed steel and plated for corrosion resistance. The coil is wound on a bobbin molded of high temperature nylon and fully shrouded. The rotor spindle is hardened and polished and runs in nylon bearings requiring no maintenance.

Fig. 18–4. Disc magnet two-phase stepper motor. *(Courtesy Stock Drive Products, Div. of Designatronics, Inc.)*

Fig. 18–5. Disc magnet two-phase stepper motor. *(Courtesy Stock Drive Products, Div. of Designatronics, Inc.)*

Fig. 18–8 shows the wiring diagram for this type of motor. The step angle is 7.5° with a 5% accuracy. It weighs 68 grams and uses 12 volts for its power source. Fig. 18–9 shows the typical characteristics.

The four-phase motor in Fig. 18–10 is two-part, with a 48-pole stator that surrounds the toroidal coil with two separate *double windings* for *uni*polar operation.

The two-phase motor is two-part, with a 48-pole stator that surrounds the toroidal coil with two separate *single windings* for *bi*polar operation.

The rotor shaft which carries the high coercivity 24-pole permanent magnet is hardended, ground and polished. The motor is fitted with maintenance-free sintered bronze bearings housed in plastic. This protects the bearings from dirt and prevents seepage of the lubricating oil by capillary effect. The step angle is 7.5°. Different motors are made to operate on 3, 6, 12, or 24 volts, and weight of the motor is 65 grams. This same type of motor is also available with a 15° step angle and weighing up to 400 grams.

Another design (Fig. 18–11) for a 7.5° permanent magnet stepping motor can run clockwise or counterclockwise, depending on the input. Holding torque is measured with two adjacent phases at the rated current and can be as much as 16 oz-in with a phase current of 0.47. This is accomplished with a weight of 250 grams or 0.55 lbs.

The hybrid 1.8° stepping motor is shown in Figs. 18–12 to 18–14.

Fig. 18–6. Disc magnet two-phase stepper motor. *(Courtesy Stock Drive Products, Div. of Designatronics, Inc.)*

Fig. 18–7. PM rotor-type step motor. *(Courtesy Stock Drive Products, Div. of Designatronics, Inc.)*

Fig. 18–8. Wiring diagram for PM rotor-type step motor. *(Courtesy Stock Drive Products, Div. of Designatronics, Inc.)*

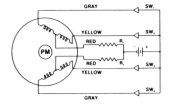

TYPICAL CHARACTERISTICS at 20°C.

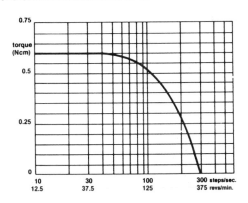

Fig. 18–9. Typical characteristics curve for PM rotor-type step motor. *(Courtesy Stock Drive Products, Div. of Designatronics, Inc.)*

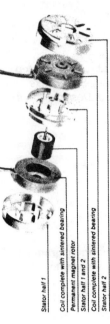

Reversing synchronous motor with three stator windings

Connection diagram for single-phase supply

Connection diagram for three-phase supply

Stator half 1

Coil complete with sintered bearing

Permanent magnet rotor

Stator half 1 and 2

Coil complete with sintered bearing

Stator half 2

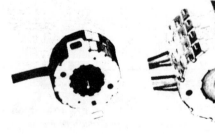

Fig. 18–10. Two-phase or four-phase, two-part, 48-pole stator motor. *(Courtesy of Stock Drive Products, Div. of Designatronics, Inc.)*

Fig. 18–11.   7.5° permanent magnet stepper motor. *(Courtesy of Stock Drive Products, Div. of Designatronics, Inc.)*

Fig. 18–12.   1.8° hybrid stepping motor. *(Courtesy of Stock Drive Products, Div. of Designatronics, Inc.)*

Fig. 18–13.   1.8° hybrid stepping motor. *(Courtesy of Stock Drive Products, Div. of Designatronics, Inc.)*

Fig. 18–14.   1.8° hybrid stepping motor. *(Courtesy of Stock Drive Products, Div. of Designatronics, Inc.)*

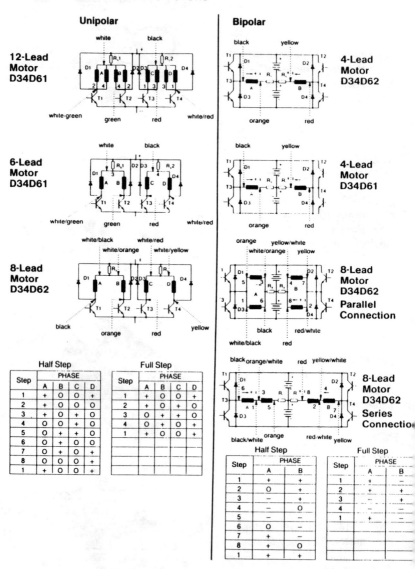

Fig. 18–15. Stepper motor wiring data. *(Courtesy of Stock Drive Products, Div. of Designatronics, Inc.)*

Fig. 18–16. Two-phase linear stepper motor. *(Courtesy of Stock Drive Products, Div. of Designatronics, Inc.)*

It can run clockwise or counter-clockwise, depending on the input and can weigh in at 340 grams (0.75 lbs) up to 2.2 lbs or 1 kg. The motors are also available with double ended shafts.

The motor shown in Fig. 18–13 weighs 10 pounds or 4.6 kg with a holding torque of 781 oz-in. It, too, is a 1.8° hybrid stepper. Fig. 18–14 is somewhat larger and weighs 23.1 lbs or 10.5 kg. Holding torque is 2130 oz-in.

As can be seen from this information, stepper motors are available in a wide variety of sizes, shapes and holding torque.

Wiring data is important for checking for proper motor operation. Fig. 18–15 shows the unipolar and bipolar connections with 6-, 8- and 12-lead motor designs. Sequencing for half-step and full-step is also given.

Two- and Four-Phase Linear Stepper Motors

The linear motor consists of a moving element called a *forcer* which travels along a special ferromagnetic track called a *platen* (Fig. 18–16). As the electromagnets which make up the forcer respond to varying currents, imbedded teeth in the forcer move to successive positions relative to similar teeth in the platen to produce the commanded motion. The motor shown in Fig. 18–16 produces a static force of 4.0 to 4.8 pounds while the motor in Fig. 18–17 is somewhat larger

Fig. 18–17. Two- and four-phase linear stepper motor. *(Courtesy of Stock Drive Products, Div. of Designatronics, Inc.)*

Fig. 18–18.   Two- and four-phase linear stepper motor. *(Courtesy Stock Drive Products, Div. of Designatronics, Inc.)*

and can produce a static force of 7.5 to 10 pounds with the two-phase unit and 8 to 11 pounds with the four-phase. This unit is suited for small to moderate size, high-performance positioning systems where it can replace rotary motors and the hardware associated with converting rotary to linear motion. The complete absence of moving contact parts provides the motor with design simplification, size reduction, unlimited life expectancy and reliability.

The four-phase has an advantage over two-phase operation since it reduces cyclic error and offers slightly higher static force due to its lower force ripple. In high speed operation, the benefits of four-phase operation are available with no increase in electronics hardware. Features of this type of motor are fine resolution, repeatability of 0.00005" (open loop) and a velocity of 100 inches per second. It also has air bearings.

For larger size systems the motor in Fig. 18–18 will do the job—it can replace many of the rotary motors and their associated hardware. This type has a static force of 44 pounds and can move at 100 ips with a repeatability of 0.00005" (open loop).

The linear motor platen (the flat part of the motor) is shown in Fig. 18–19. There are two types—the bar, or flat, type and the tubular type. The tube type weighs 3.75 pounds and the bar type is available in three different sizes: 3.5, 7, 14 pounds.

If larger sizes are required the 7300 series (Fig. 18–20) is available and it can provide much larger platens up to 72" long and weighing in at 167 pounds.

Brushless DC Motors

The brushless DC motor (Fig. 18–21) is a combination of the unirotational synchronous motor and an electronic circuit which enables it

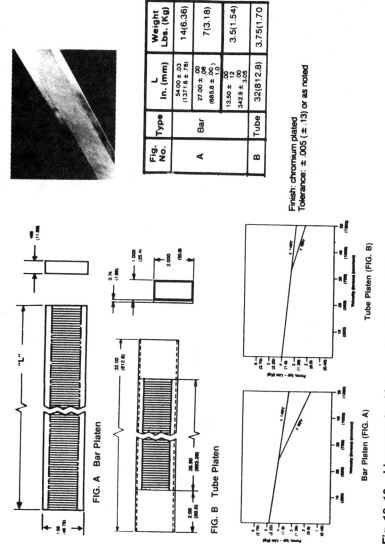

Fig. No.	Type	L In. (mm)	Weight Lbs. (Kg)
A	Bar	54.00 ±.03 (1371.6 ±.76)	14(6.36)
		27.00 ±.00/.08 (685.8 ±.00/1.0)	7(3.18)
		13.50 ±.00/.12 342.9 ±.00/3.05	3.5(1.54)
B	Tube	32(812.8)	3.75(1.70)

Finish: chromium plated
Tolerance: ± .005 (± .13) or as noted

FIG. A Bar Platen

FIG. B Tube Platen

Bar Platen (FIG. A)

Tube Platen (FIG. B)

Fig. 18–19. Linear motor platens. *(Courtesy Stock Drive Products, Div. of Designatronics, Inc.)*

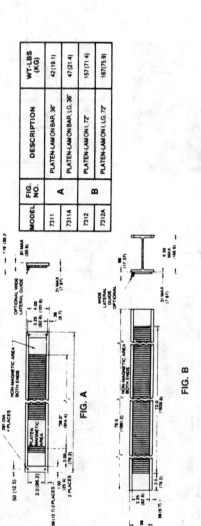

MODEL	FIG. NO.	DESCRIPTION	WT-LBS (KG)
7311	A	PLATEN-LAM ON BAR. 36"	42 (19.1)
7311A		PLATEN-LAM ON BAR. LG. 36"	47 (21.4)
7312	B	PLATEN-LAM ON I. 72"	157 (71.4)
7312A		PLATEN-LAM ON I. LG. 72"	167 (75.9)

Fig. 18–20. Larger linear motor platens. *(Courtesy Stock Drive Products, Div. of Designatronics, Inc.)*

Fig. 18–21. Brushless DC motor. *(Courtesy of Stock Drive Products, Div. of Designatronics, Inc.)*

to be connected to DC. The electronics converts the DC into the pulse form required for the synchronous motor by means of an RC oscillator. The result is a brushless DC motor having the advantages of a synchronous motor—no brush wear, long life, and the speed is not load dependent.

The three-phase brushless DC motor with Hall sensors has 12 poles and 1.4 oz-in of torque per ampere. It is small in size, only 16mm thick, and is manufactured with an internal series resistor which significantly reduces temperature tracking errors.

Ironless Rotor DC Motors

The construction of this small motor (Fig. 18–22), with its permanent magnet and ironless rotor, results in a high power-to-volume ratio, smooth torque with almost no ripple, very rapid starting, and very low electrical noise at commutation. The low inertia rotor exhibits no cogging.

The precious metal commutator (silver alloy) and brushes (gold alloy) as well as the ferrite magnet with 12 poles make a small size

Fig. 18–22. Ironless rotor DC motor. *(Courtesy of Stock Drive Products, Div. of Designatronics, Inc.)*

Fig. 18–23. Larger ironless rotor DC motor. *(Courtesy Stock Drive Products, Div. of Designatronics, Inc.)*

(8.2mm thick) motor with very low inductance and low moment of inertia. It can be obtained in 5-volt or 12-volt models. Torque of 1.05 oz-in is provided by the 5-volt motor while the 12-volt size produces 2.08 oz-in per ampere torque. A series resistor is built in for precise speed control. No-load speed is 6100 rpm and 7200 rpm for the 12-volt model.

Not all ironless rotor DC motors resemble Fig. 18–22; many resemble Fig. 18–23. These are made for operation on 15 volts or 24 volts with no-load speeds of 4300 or 5400 rpm, depending on the voltage applied. They pull 20 mA for the 15-volt model while the 24-volt pulls 25mA. The weight of the motor is 230 grams.

One of the main advantages of this type of motor over some conventional types is its ability to deliver several times the nominal stall torque, without danger of demagnetization by high current pulses.

PART SEVEN

Motor Terminology

CHAPTER 19

Motor Terminology

Ordering of parts and the selection of motors all depend upon being able to talk with or communicate with the people involved in the field every day—this chapter will enable you to become familiar with common terms. This chapter will look more like a glossary of terms than a regular chapter.*

Air Gap   The space between the rotating and stationary member in an electric motor.

Air Over (AO)   Motors intended for fan and blower service and cooled by the air stream from the fan or blower.

Alternating Current (AC)   The commonly available electric power supplied by an AC generator and distributed in one-, two-, or three-phase form. This is the standard type of power supplied to homes, businesses, industry, and farms.

Ambient   For air-cooled rotating machinery, the ambient is considered the air surrounding the motor. The temperature of the space around the motor should not exceed 40°C or 104°F.

*The Bodine Electric Company and the General Electric Company have been very helpful in furnishing information for this chapter.

Ampere   The constant current which, if maintained in two straight parallel conductors of infinite length and negligible cross section and spaced 1 meter apart in a vacuum, produces between them 2 × 10^{-7} Newton per meter of length. The unit of measurement for current. The ampere is abbreviated as A or, in some cases, amp.

Ampere Turn   The magnetomotive force produced by a current of one ampere in a coil of one turn.

Angular Velocity   Angular displacement per unit time, measured in degrees/time or radians/time.

Armature   The portion of the magnetic structure of a DC or universal motor which rotates.

Armature Reaction   The current that flows in the armature winding of a DC motor tends to produce magnetic flux in addition to that produced by the field current. This effect, which reduces the torque capacity, is called armature reaction and can affect the commutation and the magnitude of the motor's generated voltage.

Basic Speed   The speed which a motor develops at rated voltage with rated load applied.

Bearings (BRGS)   Sleeve-type bearings (SLV) are preferred where low noise level is important. The bearing resembles a short length of bronze tubing with grooves to direct oil flow (Fig. 2–4). Ball bearings are used where higher load capacity is required or periodic lubrication is impractical (Fig. 19–1).

Braking Torque   The torque required to bring a motor down from running speed to a standstill. The term is also used to describe the torque developed by a motor during dynamic braking conditions.

Breakdown Torque   The maximum torque a motor will develop at rated voltage without a relatively abrupt drop or loss in speed.

Brush   A piece of current-conducting material (usually carbon or graphite) which rides directly on the commutator of a commutated motor and conducts current from the power supply to the armature windings.

Canadian Standards Association (CSA)   Sets safety standards for motors and other electrical equipment used in Canada.

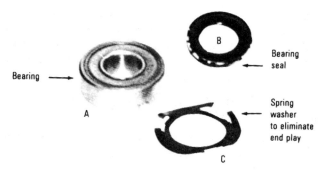

Fig. 19–1. End play is eliminated by using a spring-loaded washer (C) to hold the ball bearing (A) in place. The bearing seal (B) is used to prevent dust and dirt from getting into the bearing.

Cantilever Load   A load which tends to impose a radial force (perpendicular to the shaft axis) on a motor or gearmotor output shaft.

Capacitor   A device that, when connected in an alternating current circuit, causes the current to lead the voltage in time phase. The peak of the current wave is reached ahead of the voltage wave. This is the result of the successive storage and discharge of electric energy.

Center Ring   That part of a motor housing which supports the stator or field core.

Centrifugal Cut-Out Switch   A centrifugally-operated automatic mechanism used in conjunction with split-phase and other types of induction motors. Centrifugal cut-out switches will open or disconnect the starting winding when the rotor has reached a predetermined speed, and reconnect it when the motor speed falls below it. Without such a device, the starting winding would be susceptible to rapid overheating and subsequent burnout (Fig. 19–2).

Cogging   A term used to describe nonuniform angular velocity. It refers to rotation occurring in jerks or increments rather than smooth motion. When an armature coil enters the magnetic field pro-

Fig. 19–2. The centrifugal switch on this motor is mounted outside the frame with an arm extended into the housing to operate the switch.

duced by the field coils, it tends to speed up and slow down when leaving it. This effect becomes apparent at low speeds. The fewer the number of coils, the more noticeable it can be.

Commutator A cylindrical device mounted on the armature shaft and consisting of a number of wedge-shaped copper segments arranged around the shaft (insulated from it and each other). The motor brushes ride on the periphery of the commutator and electrically connect and switch the armature coils to the power source.

Conductor Any material which tends to make the flow of electrical current relatively easy (copper, aluminum, gold, silver, and others).

Counter Electromotive Force (CEMF) The induced voltage in a motor armature caused by conductors moving through, or "cutting," field magnetic flux. This induced voltage opposes the armature current and tends to reduce it.

Direct Current (DC) Type of power supply available from batteries or generators (not alternators) used for special-purpose applications. Current flows in one direction only; it does not alternate.

Duty Cycle The relationship between operating and rest time. A motor which can continue to operate within the temperature limits of its insulation system after it has reached normal operating (equilibrium) temperature is considered to have a continuous duty (CONT) rating. One which never reaches equilibrium temperature but is permitted to cool down between operations is operating under intermittent duty (INT) conditions.

Dynamic Unbalance A noise-producing condition caused by the nonsymmetrical weight distribution of a rotating member. The lack of uniform wire spacing in a wound armature or casting voids in a rotor or fan assembly can cause relatively high degrees of unbalance.

Eddy Current Localized currents induced in an iron core by alternating magnetic flux. These currents translate into losses (heat), and their minimization is an important factor in lamination design.

Efficiency The efficiency of a motor is the ratio of mechanical output to electrical input. It represents the effectiveness with which the motor converts electrical energy into mechanical energy.

Electrical Coupling When two coils are so situated that some of the flux set up by either coil links some of the turns of the other, they are said to be electrically coupled.

Electromotive Force (EMF) A synonym for voltage, usually restricted to generated voltage.

Encapsulated Winding A motor that has its winding structure completely coated with an insulating resin (such as epoxy). This construction type is more designed for exposure to severe atmospheric conditions than is the normal varnished winding.

Enclosure (ENCL) The term used to describe the motor housing. Some of the more common types are:

Drip-Proof (DP)   Ventilation openings in end shields and shells placed so drops of liquid falling within an angle of 15° from vertical will not affect performance. Usually used indoors, in fairly clean, dry locations.

Explosion-Proof (EXP-PRF)   A special enclosed motor designed to withstand an internal explosion of specified gases or vapors and allow the internal flame or explosion to escape. Usually used in smaller ratings below ⅓ horsepower if nonventilated (EPNV) and in fan-cooled (EPFC) in larger ratings.

Fan-Cooled (TEFC)   Includes an integral fan to blow cooling air over the motor.

Nonventilated (TENV)   Not equipped with a fan for external cooling. Depends on convection air for cooling.

Open (OP)   Ventilation openings in end shields and/or shell to permit passage of cooling air over and around the windings.

Fig. 19–3.   An open-enclosure motor (OP).

Locations of openings not restricted. For use indoors, in fairly clean locations (Fig. 19–3).

Totally Enclosed (TE)   No openings in the motor housing (but not airtight). Used in locations which are dirty, oily, and the like. The two types are:

End Shield   That part of the motor housing which supports the bearing and acts as a protective guard to the electrical and rotating parts inside the motor. This part is frequently called the end bracket or end belt.

Excitation Current   A term applied to the current in the shunt field of a motor resulting from voltage applied across the field.

Farad   A unit of measurement for electrical capacitance. A capacitor has a capacitance of 1 farad when a potential difference of 1 volt will charge it with 1 coulomb of electricity.

Feedback   As it generally relates to motors and controls, feedback refers to the voltage information received by a feedback circuit. Depending on a predetermined potentiometer setting, a motor control can correct the voltage to deliver appropriate speed and/or torque.

Field   A term commonly used to describe the stationary (stator) member of a DC motor. The field provides the magnetic field with which the mechanically-rotating (armature) member interacts.

Field Weakening   The introduction of resistance in series with the shunt-wound field of a motor to reduce the voltage and current that weakens the strength of the magnetic field and thereby increases the motor speed.

Flux   The magnetic field that is established around an energized conductor or permanent magnet.

Form Factor   A figure of merit that indicates how much rectified current departs from pure (nonpulsating) DC. A large departure from form factor (pure DC) increases the heating effect of the motor and reduces brush life.

Fractional-Horsepower Motor   A motor with continuous rating of less than 1 horsepower, open construction at 1700–1800 rpm.

Frame (FR)   Usually refers to the NEMA system of standardization of motor-mounting dimensions.

Frequency   The rate at which alternating current reverses its direction of flow. Measured in hertz (Hz).

Full-Load Current   The current drawn from the line when the motor is operating at full-load torque and full-load speed at rated frequency and voltage.

Full-Load Torque   The torque necessary to produce rated horsepower at full-load speed.

Galvanometer   An extremely sensitive instrument used to measure small current and voltage in an electrical circuit.

Gearhead   The portion of a gearmotor which contains the actual gearing that converts the basic motor speed to the rated output speed.

Horsepower (hp)   Power rating of the motor. It takes 746 watts of electrical energy to produce 1 horsepower.

Hysteresis Loss   The resistance offered by materials to becoming magnetized. Reduced by using silicon steel laminations.

Impedance   The vectorial sum of both resistance and reactance in a motor. Total opposition to current flow. Measured in ohms. Z is the impedance symbol.

Inductance   The characteristic of a coil of wire to cause the current to lag the voltage in time phase. L is the symbol for inductance. Inductance is measured in henrys (H).

Inertial Load   A load (flywheel, fan, or the like) which tends to cause the motor shaft to continue to rotate after the power has been removed.

Insulation (INSUL)   In motors, usually classified by maximum allowable operating temperature:

Class A, 105°C or 221°F
Class B, 130°C or 266°F
Class F, 155°C or 311°F
Class H, 180°C or 365°F

Insulator   A material which tends to resist the flow of electric current.

Integral Horsepower Motor   In terms of horsepower, a motor built in a frame having continuous rating of 1 horsepower or more, open construction at 1700–1800 rpm. In terms of size, an integral horsepower motor is usually greater than 9 inches in diameter, although it can be as small as 6 inches.

Line Voltage   Voltage supplied by the power company or voltage supplied as input to the device.

Locked-Rotor Current   Steady-state current taken from the line with the rotor at standstill. Steady-state current means the current does not vary in its intensity, but remains constant.

Locked-Rotor Torque   The minimum torque that a motor will develop at rest for all angular positions of the rotor.

Magnetomotive Force (MMF)   The magnetic energy supplied with the establishment of flux between the poles of a magnet.

Mechanical Degree   The popular physical understanding of degrees (360° = 1 rotation).

Motor Types   Classified by operating characteristics and/or type of power required (Table 19–1). *Induction Motors* include single-phase and three-phase motors. Direct-current motors are further classified as *shunt, series,* and *compound.*

National Electrical Code (NEC)   A code for the purpose of practical safeguarding of persons and property from the hazards arising from the use of electricity. It is sponsored by the National Fire Protection Institute, and is used to serve as a guide for governmental bodies whose duty is to regulate building codes.

NEMA   The National Electrical Manufacturers Association. This organization establishes certain voluntary industry standards relating to motors. These standards refer to the operating characteristics, terminology, basic dimensions, ratings, and testing.

Open Circuit   An open circuit in a motor is a defect that causes an interruption in the path through which the electric current normally flows.

Table 19–1. Motor Characteristics

		Duty	Typical Reversibility	Speed Character	Typical Start Torque*
POLYPHASE	AC	Continuous	Rest/Rot.	Relatively Constant	175% & up
SPLIT PHASE Synchronous	AC	Continuous	Rest Only	Relatively Constant	125–200%
SPLIT PHASE Nonsynchronous	AC	Continuous	Rest Only	Relatively Constant	175% & up
PSC Nonsynchronous High slip	AC	Continuous	Rest/Rot.†	Varying	175% & up
PSC Nonsynchronous Normal slip	AC	Continuous	Rest/Rot.†	Relatively Constant	75–150%
PSC Reluctance Synchronous	AC	Continuous	Rest/Rot.†	Constant	125–200%
PSC Hysteresis Synchronous	AC	Continuous	Rest/Rot.†	Constant	125–200%
SHADED POLE	AC	Continuous	Uni-Directional	Constant	75–150%
SERIES	AC/DC	Int./Cont.	Uni-Directional•	Varying‡	175% & up
PERMANENT MAGNET	DC	Continuous	Rest/Rot.§	Adjustable	175% & up
SHUNT	DC	Continuous	Rest/Rot.	Adjustable	125–200%
COMPOUND	DC	Continuous	Rest/Rot.	Adjustable	175% & up
SHELL ARM	DC	Continuous	Rest/Rot.	Adjustable	175% & up
PRINTED CIRCUIT	DC	Continuous	Rest/Rot.	Adjustable	175% & up
BRUSHLESS DC	DC	Continuous	Rest/Rot.	Adjustable	75–150%
DC STEPPER	DC	Continuous	Rest/Rot.	Adjustable	■

*Percentages are relative to full-load rated torque. Categorizations are general and apply to small motors.
■Dependent upon load inertia and electronic driving circuitry.
•Usually unidirectional—can be manufactured bidirectional.
†Reversible while rotating under favorable conditions: generally when inertia of the driven load is not excessive.
‡Can be adjusted, but varies with load.
§Reversible down to 0°C after passing through rest.

Phase   A term that indicates the space relationship of windings and changing values of the recurring cycles of AC.

Phase Displacement   Mechanical and electrical angle by which phases in a polyphase motor or main and capacitor (or starting) windings in an induction motor are displaced from one another.

Plug Reversal   Reconnecting a motor's windings to reverse its direction of rotation while running.

Polarities   Terms (positive, negative, north, and south) that indicate the direction of current and flux flow in electrical and magnetic circuits at any given instant.

Power Factor (PF)   A measurement of the time-phase difference between the voltage and current in an AC circuit. It is represented by the cosine of the angle of the phase difference. Zero degrees has a power factor of 100 percent. That means the watts and volt-amperes are equal and there is nothing more than resistance in the circuit. Ninety degrees of angle represents nothing in the way of resistance and only inductance in the circuit. PF is also found by the formula:

$$\frac{\text{True Power (TP)}}{\text{Apparent Power (AP)}}$$

Prony Brake   A simple mechanical device, normally made of wood with an adjustable leather strap, that is used to test for the torque output of a motor. The prony brake loads the motor and a spring scale attached to it gives a relatively accurate measurement or torque.

Pull-In Torque   The maximum constant torque that a synchronous motor will accelerate into synchronism at rated voltage and frequency.

Pull-Up Torque   The minimum torque delivered by an AC motor during the period of acceleration from zero to the speed at which breakdown occurs. For motors which do not have a definite breakdown torque, the pull-up torque is the minimum torque developed during the process of getting up to rated speed.

Rectifier   An electronic circuit which converts alternating current into direct current.

Reluctance   The characteristic of a magnetic material which resists the flow of magnetic lines of force through it.

Resilient Mounting   A suspension system or cushioned mounting designed to reduce the transmission of normal motor noise and vibration to the mounting surface.

Resistance   The degree of obstacle presented by a material to the flow of electrical current is known as resistance. Resistance is measured in ohms. R is the symbol for resistance.

Rotor   The rotating member of an induction motor in a single-phase device. Current that is normally induced in the rotor reacts with

Fig. 19–4.   This single-phase motor rotor has the centrifugal switch mechanism mounted with springs holding it in place. Note the three pressure pads on that part of the mechanism that moves outward to operate the switch.

the magnetic field produced by the stator. This produces torque and rotation (Fig. 19–4).

Salient Pole   A motor has salient poles when its stator or field poles are concentrated into confined arcs and the winding is wrapped around them (as opposed to distributing them in a series of slots).

Secondary Winding   The secondary winding of a motor is a winding that is not connected to the power source but carries current induced in it through its magnetic linkage with the primary winding.

Semiconductor   A material, usually silicon or germanium, that permits limited current flow.

Service Factor (SF)   A measure of the overload capacity designed into a motor. A 1.15 SF means the motor can deliver 15 percent more than the rated horsepower without injurious overheating. A 1.0 SF motor should not be overloaded beyond its rated horsepower. Service factors will vary for different horsepower motors and for different speeds. Standard NEMA service factors for various horsepower motors and motor speeds are shown in Table 19–2, below, for easy reference.

Short Circuit   A defect in a winding that causes part of the normal electrical circuit to be bypassed.

Skew   Arrangement of laminations on a rotor or armature to provide a slight diagonal pattern of their slots with respect to the shaft axis.

Table 19–2. Service Factors

Horsepower	Service Factor			
	Synchronous Speed			
	3600	1800	1200	900
$\frac{1}{20}, \frac{1}{12}, \frac{1}{8}$	1.40	1.40	1.40	1.40
$\frac{1}{6}, \frac{1}{4}, \frac{1}{3}$	1.35	1.35	1.35	1.35
$\frac{1}{2}$	1.25	1.25	1.25	1.15
$\frac{3}{4}$	1.25	1.25	1.15	1.15
1	1.25	1.15	1.15	1.15
$1\frac{1}{2}$ and up	1.15	1.15	1.15	1.15

This pattern helps to eliminate low-speed cogging effects in an armature and minimize induced vibration in a rotor.

Slip   The difference between the speed of the rotating magnetic field (which is always synchronous) and the rotor in a nonsynchronous induction motor. Slip is expressed as a percentage of a synchronous speed and generally increases with an increase in load.

Slip Ring   A conductor band, mounted on an armature and insulated from it. A conductor strip slides on the band as the armature rotates. The function of the slip-ring system is essentially the same as a commutator and brushes. Slip rings are also used to transmit current from the armature in a generator application.

Starting Torque   The torque or twisting force delivered by a motor when energized.

Stator   That part of an induction motor's magnetic structure that does not rotate. It usually contains the primary winding.

Synchronous Speed   The speed of the rotating magnetic field set up by an energized stator winding. In synchronous motors, the rotor locks into synchronism with the field, and is said to run at synchronous speed.

Tachometer   A small generator normally used as a velocity-sensing device. Tachometers are typically attached to the output shaft of DC servo motors requiring close speed regulation. The tachometer feeds its signal to a control which adjusts its output to the DC motor accordingly (called *closed loop feedback* control).

Temperature Rise   The amount by which a motor, operating under rated conditions, is hotter than its surrounding ambient temperature.

Thermal Protector   A protective device, built into the motor, that disconnects the motor from its power source if the temperature becomes excessive for any reason.

Thermocouple   A junction of two dissimilar materials which generates a minute voltage in proportion to its temperature. Such devices may be used as signal source in indicating instruments and control equipment.

Torque   Turning force delivered by a motor or gearmotor shaft, usually expressed in ounce-inches or Newton-meters. There are three types of torque associated with electric motors: *starting torque, full-load torque,* and *breakdown torque.*

Underwriters Laboratories, Inc. (UL)   An independent testing organization that sets safety standards for motors and other electrical equipment.

Variable Resistor   A resistor, connected in series with a motor, that can be adjusted to vary the amount of current available and thereby alter motor speed.

Voltage   The force that causes a current to flow in an electrical circuit. Analogous to pressure in hydraulics, voltage is often referred to as electrical pressure. Voltage is measured in volts (V).

Voltage Drop   Loss encountered across a circuit impedance. Voltage drop across a resistor takes the form of heat released into the air at the point of resistance.

Watt   The amount of power required to maintain a current of 1 ampere at a pressure of 1 volt. One horsepower is equal to 746 watts. The symbol for watt is W.

Appendix

Fractional-Horsepower Motors and Their Applications

Applications	Motor Diameter, Inches								
	2½ × 3	3.4			3.8	4.9		5.6	
	33&44 Frame	51 Frame	69 Frame	11 Frame	19 Frame	29 Frame	39 Frame	Form G2(GE)
Refrigerators & Freezers	●								●
Commercial Refrigeration Vending machines, display cases, unit coolers, water coolers, condensing packages, carbonator pumps	●	●	●	●		●	●		
Room Air Conditioners Portables, window and casement units, thru-the-wall units			●			●	●		
Central Heat & Air Conditioning Furnaces, outdoor condensers heat pumps, blower packages						●	●		●
Fans & Vents Window and floor fans, attic and room vents, kitchen and bathroom vents, range hoods	●		●		●	●			●
Pumps & Blowers Sump pumps, swimming pool pumps, commercial and industrial blowers			●		●	●			●
Other Home Shoe polisher, juicer, compact washer, garage-door openers, health equipment, microwave ovens	●	●			●				
Other Specialty Markets Room heaters, hot-water circulators, oil burners, fan coil units, business machines, hand dryers, tools	●	●	●	●		●	●		●

NEMA Motor Frame Dimension Standards

Standardized motor dimensions as established by the National Electrical Manufacturers Association (NEMA) are tabulated below and apply to all base-mounted motors listed herein which carry a NEMA frame designation.

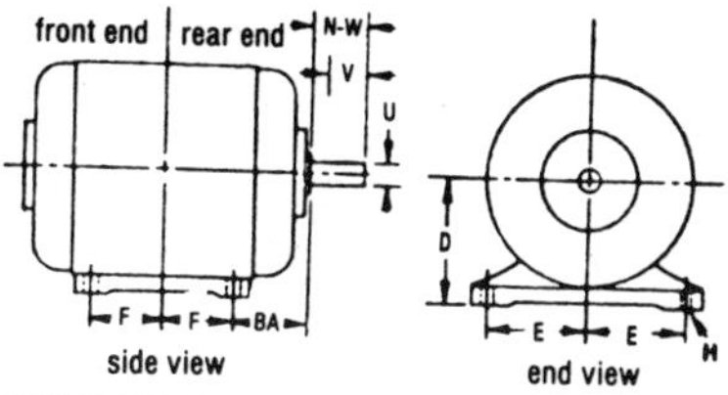

NEMA Frame	D*	2E	2F	BA	H	N-W	U	V§ Min.	Key Wide	Key Thick	Key Long	NEMA Frame
	—All Dimensions in Inches—											
42	2 5/8	3 1/2	1 11/16	2 1/16	9/32 slot	1 1/8	3/8	—	—	3/64 flat	—	42
48	3	4 1/4	2 3/4	2 1/2	11/32 slot	1 1/2	1/2	—	—	3/64 flat	—	48
56 56H	3 1/2	4 7/8	3 3&5‡	2 3/4	11/32 slot	1 7/8†	5/8†	—	3/16†	3/16†	1 3/8†	56 56H
56HZ	3 1/2	**	**	**	**	2 1/4	7/8	2	3/16	3/16	1 3/8	56Hz
66	4 1/8	5 7/8	5	3 1/8	13/32 slot	2 1/4	3/4	—	3/16	3/16	1 7/8	66
143T 145T	3 1/2	5 1/2	4 5	2 1/4	11/32 dia.	2 1/4	7/8	2	3/16	3/16	1 3/8	143T 145T
182 184	4 1/2	7 1/2	4 1/2 5 1/2	2 3/4	13/32 dia.	2 1/4	7/8	2	3/16	3/16	1 3/8	182 184
182T 184T			4 1/2 5 1/2			2 3/4	1 1/8	2 1/2	1/4	1/4	1 3/4	182T 184T
213 215	5 1/4	8 1/2	5 1/2 7	3 1/2	13/32 dia.	3	1 1/8	2 3/4	1/4	1/4	2	213 215
213T 215T			5 1/2 7			3 3/8	1 3/8	3 1/8	5/16	5/16	2 3/8	213T 215T
254U 256U	6 1/4	10	8 1/4 10	4 1/4	17/32 dia.	3 3/4	1 3/8	3 1/2	5/16	5/16	2 3/4	254U 256U
254T 256T			8 1/4 10			4	1 5/8	3 3/4	3/8	3/8	2 7/8	254T 256T
284U 286U	7	11	9 1/2 11	4 3/4	17/32 dia.	4 7/8	1 5/8	4 5/8	3/8	3/8	3 3/4	284U 286U
284T 286T			9 1/2 11			4 5/8	1 7/8	4 3/4	1/2	1/2	3 1/4	284T 286T
324U 326U	8	12 1/2	10 1/2 12	5 1/4	21/32 dia.	5 3/8	1 7/8	5 5/8	1/2	1/2	4 1/4	324U 326U
324T 326T			10 1/2 12			5 1/4	2 1/8	5	1/2	1/2	3 7/8	324T 326T
326TS			12			3 3/4■	1 7/8■	3 1/2■	1/2	1/2	2■	326TS
364U 365U	9	14	11 1/4 12 1/4	5 7/8	21/32 dia.	6 3/8	2 1/8	6 1/8	1/2	1/2	5	364U 365U

* Dimension D will never be greater than the above values on rigid mount motors, but it may be less so that shims up to 1/32" thick (1/16" on 364U and 365U frames) may be required for coupled or geared machines.

‡ Dayton motors designated 56H have two sets of 2F mounting holes—3" and 5".

■ Standard short shaft for direct-drive applications.

** Base of the 56HZ frame motors has holes and slots to match NEMA 56, 56H, 143T and 145T mounting dimensions.

† Certain NEMA 56Z frame motors have 1/2" dia. ω 1 1/2" long shaft with 3/64" flat. These exceptions are noted in this catalog.

§ Dimension V is shaft length available for coupling, pinion, or pulley hub—this is a minimum value.

NEMA Letter Designations Following Frame Number

C—Face mount.
H—Has 2F dimension larger than same frame without H suffix.
J—Face mount to fit jet pumps.
K—Has hub for sump pump mounting.
M—Flange mount for oil burner.
N—Flange mount for oil burner.
T,U—Integral HP motor dimension standards set by NEMA in 1964 and 1953, respectively.
Y—Non-standard mounting; see manufacturer's drawing for mounting dimensions.
Z—Non-standard shaft extension (NW,U dimensions).
For their own identification, manufacturers may use a letter before the NEMA frame number. This has no reference to mounting dimensions.

(Courtesy Bodine)

Motor Schematics

AC Single-Phase

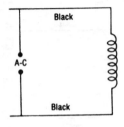

SHADED POLE. Nonreversible.

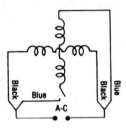

SPLIT-PHASE: To reverse (from rest only): transpose blue leads or black leads.

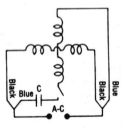

CAPACITOR-START. To reverse (from rest only): transpose blue leads or black leads.

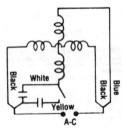

TWO-VALVE CAPACITOR. Two-capacitor start, one-capacitor run. To reverse (from rest only): transpose the black leads.

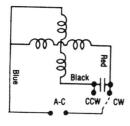

PERMANENT-SPLIT CAPACITOR (3-LEAD). To reverse: connect either side of capacitor to line.

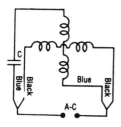

PERMANENT-SPLIT CAPACITOR (4-LEAD). To reverse: transpose the black leads.

AC Polyphase

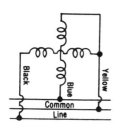

TWO-PHASE (3-WIRE). To reverse: transpose blue and black leads.

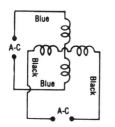

TWO-PHASE (4-LEAD). To reverse: transpose blue or black leads.

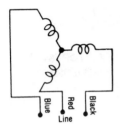

THREE-PHASE (SINGLE VOLTAGE).
To reverse: transpose any two leads.

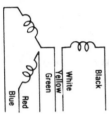

THREE-PHASE (STAR OR DELTA).
For 440V—connect together white, yellow, and green. Connect to line black,
red, and blue. To reverse: transpose
any two line leads. For 220V—connect
white to blue, black to green, and yellow to red. Then connect each junction
point to line. To reverse: transpose
any two junction points with line.

DC

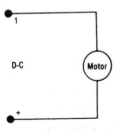

**PERMANENT-MAGNET, BRUSHLESS
DC, PRINTED CIRCUIT, SHELL-TYPE
ARMATURE.** To reverse: transpose
motor leads.

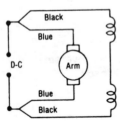

SHUNT WOUND. To reverse: transpose blue or black leads.

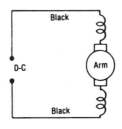

SERIES WOUND (2-LEAD). Non-reversible.

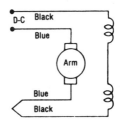

SERIES WOUND (4-LEAD). To reverse: transpose blue leads.

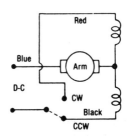

SERIES WOUND (SPLIT-FIELD). To reverse: connect other field lead to line.

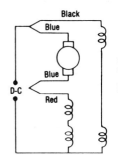

COMPOUND WOUND (5-WIRE REVERSIBLE). To reverse: transpose blue leads.

Formulas for Motor Applications

T = torque or twisting moment (Force × moment arm length)
π = 3.1416
N = revolutions per minute
HP = horsepower (33,000 ft.-lbs. per min.) applies to power output
R = radius of pulley, in feet
E = input voltage
I = current in amperes
P = power input in watts

$$HP = \frac{T\,(\text{lb-in.}) \times N\,(\text{rpm})}{63{,}025}$$

$$HP = T\,(\text{oz-in.}) \times N \times 9.917 \times 10^{-7}$$
$$= \text{approx. } T\,(\text{oz-in.}) \times N \times 10^{-6}$$

$$P = EI \times \text{power factor} = \frac{HP \times 746}{\text{motor efficiency}}$$

Power to Drive Pumps:

$$HP = \frac{\text{Gal. per min.} \times \text{Total Head (inc. friction)}}{3{,}960 \times \text{eff. of pump}}$$

Where:

Approx. friction head (ft.) =

$$\frac{\text{pipe length (ft.)} \times [\text{velocity of flow (fps)}]^2 \times 0.02}{5.367 \times \text{diameter (in.)}}$$

Eff. = Approx. 0.50 to 0.85

Time to Change Speed of Rotating Mass:

$$\text{Time (sec.)} = \frac{WR^2 \times \text{change in rpm}}{308 \times \text{torque (ft-lb.)}}$$

$$\text{Where: } WR^2 \text{ (disc)} = \frac{\text{Weight (lbs.)} \times [\text{radius (ft.)}]^2}{2}$$

WR^2 (rim) =

$$\frac{\text{Wt. (lbs.)} \times [(\text{outer radius in ft.})^2 + \text{inner radius in ft.})^2]}{2}$$

Power to Drive Fans:

$$HP = \frac{\text{Cu. ft. air per min.} \times \text{water gage pressure (in.)}}{6.350 \times \text{Eff.}}$$

Horsepower/Watts vs. Torque Conversion Chart

hp	watts	@ 1125 rpm Oz.-in.	@ 1125 rpm mN·m	@ 1200 rpm Oz.-in.	@ 1200 rpm mN·m	@ 1425 rpm Oz.-in.	@ 1425 rpm mN·m
1/2000	0.373	0.4482	3.1649	0.4202	2.9670	0.3538	2.4986
1/1500	0.497	0.5976	4.2198	0.5602	3.9561	0.4718	3.3314
1/1000	0.746	0.8964	6.3297	0.8403	5.9341	0.7077	4.9971
1/750	0.994	1.1951	8.4396	1.1205	7.9121	0.9435	6.6628
1/500	1.49	1.7927	12.6594	1.6807	11.8682	1.4153	9.9943
1/200	3.73	4.4818	31.6485	4.2017	29.6705	3.5383	24.9857
1/150	4.97	5.9757	42.1980	5.6023	39.5606	4.7177	33.3142
1/100	7.46	8.9636	63.2970	8.4034	59.3409	7.0765	49.9713
1/75	9.94	11.9515	84.3960	11.2045	79.1212	9.4354	66.6284
1/70	10.70	12.8052	90.4243	12.0048	84.7727	10.1093	71.3876
1/60	12.40	14.9393	105.4950	14.0056	98.9015	11.7942	83.2855
1/50	14.90	17.9272	126.5940	16.8068	118.6818	14.1531	99.9426
1/40	18.60	22.4090	158.2425	21.0085	148.3523	17.6913	124.9283
1/30	24.90	29.8787	210.9899	28.0113	197.8031	23.5884	166.5710
1/25	29.80	35.8544	253.1879	33.6135	237.3637	28.3061	199.8852
1/20	37.30	44.8180	316.4849	42.0169	296.7046	35.3827	249.8565
1/15	49.70	59.7574	421.9799	56.0225	395.6061	47.1769	333.1420
1/12	62.10	74.6967	527.4748	70.0282	494.5077	58.9711	416.4275
1/10	74.6	89.6361	632.9698	84.0338	593.4092	70.7653	499.7130
1/8	93.2	112.0451	791.2123	105.0423	741.7615	88.4566	624.6413
1/6	124.0	149.3934	1054.9497	140.0563	989.0153	117.9422	832.8550
1/4	186.0	224.0902	1582.4245	210.0845	1483.5230	176.9133	1249.2825
1/3	249.0	298.7869	2109.8994	280.1127	1978.0307	235.8844	1665.7101

hp	watts	@ 1500 rpm Oz.-in.	@ 1500 rpm mN·m	@ 1725 rpm Oz.-in.	@ 1725 rpm mN·m	@ 1800 rpm Oz.-in.	@ 1800 rpm mN·m
1/2000	0.373	0.3361	2.3736	0.2923	2.0640	0.2801	1.9780
1/1500	0.497	0.4482	3.1648	0.3897	2.7520	0.3735	2.6374
1/1000	0.746	0.6723	4.7473	0.5846	4.1281	0.5602	3.9561
1/750	0.994	0.8964	6.3297	0.7794	5.5041	0.7470	5.2747
1/500	1.490	1.3445	9.4945	1.1692	8.2561	1.1205	7.9121
1/200	3.730	3.3614	23.7364	2.9229	20.6403	2.8011	19.7803
1/150	4.97	4.4818	31.6485	3.8972	27.5204	3.7348	26.3737
1/100	7.46	6.7227	47.4727	5.8458	41.2806	5.6023	39.5606
1/75	9.94	8.9636	63.2970	7.7944	55.0409	7.4697	52.7475
1/70	10.70	9.6039	67.8182	8.3512	58.9723	8.0032	56.5152
1/60	12.40	11.2045	79.1212	9.7431	68.8011	9.3371	65.9344
1/50	14.90	13.4454	94.9455	11.6917	82.5613	11.2045	79.1212
1/40	18.60	16.8068	118.6818	14.6146	103.2016	14.0056	98.9015
1/30	24.90	22.4090	158.2425	19.4861	137.6021	18.6742	131.8687
1/25	29.80	26.8908	185.8909	23.3833	165.1226	22.4090	158.2425
1/20	37.3	33.6135	237.3637	29.2292	206.4032	28.0113	197.8031
1/15	49.7	44.8180	316.4849	38.9722	275.2043	37.3484	263.7374
1/12	62.1	56.0225	395.6061	48.7153	344.0053	46.6854	329.6718
1/10	74.6	67.2270	474.7274	58.4583	412.8064	56.0225	395.6061
1/8	93.2	84.0338	593.4092	73.0729	516.0080	70.0282	494.5077
1/6	124.0	112.0451	791.2123	97.4305	688.0107	93.3709	659.3436
1/4	186.0	168.0676	1186.8184	146.1458	1032.0160	140.0563	989.0153
1/3	249.0	224.0902	1582.4245	194.8610	1376.0213	186.7418	1318.6871

(Courtesy Bodine)

Horsepower/Watts vs. Torque Conversion Chart (Cont'd)

hp	watts	@ 3000 rpm Oz.-in.	mN·m	@ 3450 rpm Oz.-in.	mN·m	@ 3600 rpm Oz.-in.	mN·m
1/2000	0.373	0.1681	1.1868	0.1461	1.0320	0.1401	0.9890
1/1500	0.497	0.2241	1.5824	0.1949	1.3760	0.1867	1.3187
1/1000	0.746	0.3361	2.3736	0.2923	2.0640	0.2801	1.9780
1/750	0.994	0.4482	3.1648	0.3897	2.7520	0.3735	2.6374
1/500	1.490	0.6723	4.7473	0.5846	4.1281	0.5602	3.9561
1/200	3.730	1.6807	11.8682	1.4615	10.3202	1.4006	9.8902
1/150	4.97	2.2409	15.8242	1.9486	13.7602	1.8674	13.1869
1/100	7.46	3.3614	23.7364	2.9229	20.6403	2.8011	19.7803
1/75	9.94	4.4818	31.6485	3.8972	27.5204	3.7348	26.3737
1/70	10.70	4.8019	33.9091	4.1756	29.4862	4.0016	28.2576
1/60	12.40	5.6023	39.5606	4.8715	34.4005	4.6685	32.9672
1/50	14.90	6.7227	47.4727	5.8458	41.2806	5.6023	39.5606
1/40	18.6	8.4034	59.3409	7.3073	51.6008	7.0028	49.4508
1/30	24.9	11.2045	79.1212	9.7431	68.8011	9.3371	65.9344
1/25	29.8	13.4454	94.9455	11.6917	82.5613	11.2045	79.1212
1/20	37.3	16.8068	118.6818	14.6146	103.2016	14.0056	98.9015
1/15	49.7	22.4090	158.2425	19.4861	137.6021	18.6742	131.8687
1/12	62.1	28.0113	197.8031	24.3576	172.0027	23.3427	164.8359
1/10	74.6	33.6135	237.3637	29.2292	206.4032	28.0113	197.8031
1/8	93.2	42.0169	296.7046	36.5364	258.0040	35.0141	247.2538
1/6	124.0	56.0225	395.6061	48.7153	344.0053	46.6854	329.6718
1/4	186.0	84.0338	593.4092	73.0729	516.0080	70.0282	494.5077
1/3	249.0	112.0451	791.2123	97.4305	688.0107	93.3709	659.3436

hp	watts	@ 5000 rpm Oz.-in.	mN·m	@ 7500 rpm Oz.-in.	mN·m	@ 10,000 rpm Oz.-in.	mN·m
1/2000	0.373	0.1008	0.7121	0.0672	0.4747	0.0504	0.3560
1/1500	0.497	0.1345	0.9495	0.0896	0.6330	0.0672	0.4747
1/1000	0.746	0.2017	1.4242	0.1345	0.9495	0.1008	0.7121
1/750	0.994	0.2689	1.8989	0.1793	1.2659	0.1345	0.9495
1/500	1.490	0.4034	2.8484	0.2689	1.8989	0.2017	1.4242
1/200	3.730	1.0084	7.1209	0.6723	4.7473	0.5042	3.5605
1/150	4.97	1.3445	9.4945	0.8964	6.3297	0.6723	4.7473
1/100	7.46	2.0168	14.2418	1.3445	9.4945	1.0084	7.1209
1/75	9.94	2.6891	18.9891	1.7927	12.6594	1.3445	9.4945
1/70	10.70	2.8812	20.3455	1.9208	13.5636	1.4406	10.1727
1/60	12.40	3.3614	23.7364	2.2409	15.8242	1.6807	11.8682
1/50	14.90	4.0336	28.4836	2.6891	18.9891	2.0168	14.2418
1/40	18.60	5.0420	35.6046	3.3614	23.7364	2.5210	17.8023
1/30	24.90	6.7227	47.4727	4.4818	31.6485	3.3614	23.7364
1/25	29.80	8.0672	56.9673	5.3782	37.9782	4.0336	28.4836
1/20	37.30	10.0841	71.2091	6.7227	47.4727	5.0420	35.6046
1/15	49.70	13.4454	94.9455	8.9636	63.2970	6.7227	47.4727
1/12	62.10	16.8068	118.6818	11.2045	79.1212	8.4034	59.3409
1/10	74.6	20.1681	142.4182	13.4454	94.9455	10.0841	71.2091
1/8	93.2	25.2101	178.0228	16.8068	118.6818	12.6051	89.0114
1/6	124.0	33.6135	237.3637	22.4090	158.2425	16.8068	118.6818
1/4	186.0	50.4203	356.0455	33.6135	237.3637	25.2101	178.0228
1/3	249.0	67.2270	474.7274	44.8180	316.4849	33.6135	237.3637

$mN\epsilon m = N\epsilon m\omega 10^{-3}$

Temperature Conversion Table
°F ⇄ °C

The numbers in italics in the center column refer to the temperature, either in Celsius or Fahrenheit, which is to be converted to the other scale. If converting Fahrenheit to Celsius, the equivalent temperature will be found in the left column. If converting Celsius to Fahrenheit, the equivalent temperature will be found in the column on the right.

-100 TO 30			31 TO 71			72 TO 212			213 TO 620			621 TO 1000		
C		F	C		F	C		F	C		F	C		F
-73	*(-)100*	-148	-0.6	*31*	87.8	22.2	*72*	161.6	104	*220*	428	332	*630*	1166
-68	*(-)90*	-130	0	*32*	89.6	22.8	*73*	163.4	110	*230*	446	338	*640*	1184
-62	*(-)80*	-112	0.6	*33*	91.4	23.3	*74*	165.2	116	*240*	464	343	*650*	1202
-57	*(-)70*	-94	1.1	*34*	93.2	23.9	*75*	167.0	121	*250*	482	349	*660*	1220
-51	*(-)60*	-76	1.7	*35*	95.0	24.4	*76*	168.8	127	*260*	500	354	*670*	1238
-46	*(-)50*	-58	2.2	*36*	96.8	25.0	*77*	170.6	132	*270*	518	360	*680*	1256
-40	*(-)40*	-40	2.8	*37*	98.6	25.6	*78*	172.4	138	*280*	536	366	*690*	1274
-34.4	*(-)30*	-22	3.3	*38*	100.4	26.1	*79*	174.2	143	*290*	554	371	*700*	1292
-28.9	*(-)20*	-4	3.9	*39*	102.2	26.7	*80*	176.0	149	*300*	572	377	*710*	1310
-23.3	*(-)10*	14	4.4	*40*	104.0	27.2	*81*	177.8	154	*310*	590	382	*720*	1328
-17.8	*0*	32	5.0	*41*	105.8	27.8	*82*	179.6	160	*320*	608	388	*730*	1346
-17.2	*1*	33.8	5.6	*42*	107.6	28.3	*83*	181.4	166	*330*	626	393	*740*	1364
-16.7	*2*	35.6	6.1	*43*	109.4	28.9	*84*	183.2	171	*340*	644	399	*750*	1382
-16.1	*3*	37.4	6.7	*44*	111.2	29.4	*85*	185.0	177	*350*	662	404	*760*	1400
-15.6	*4*	39.2	7.2	*45*	113.0	30.0	*86*	186.8	182	*360*	680	410	*770*	1418
-15.0	*5*	41.0	7.8	*46*	114.8	30.6	*87*	188.6	188	*370*	698	416	*780*	1436
-14.4	*6*	42.8	8.3	*47*	116.6	31.1	*88*	190.4	193	*380*	716	421	*790*	1454
-13.9	*7*	44.6	8.9	*48*	118.4	31.7	*89*	192.2	199	*390*	734	427	*800*	1472
-13.3	*8*	46.4	9.4	*49*	120.0	32.2	*90*	194.0	204	*400*	752	432	*810*	1490
-12.8	*9*	48.2	10.0	*50*	122.0	32.8	*91*	195.8	210	*410*	770	438	*820*	1508
-12.2	*10*	50.0	10.6	*51*	123.8	33.3	*92*	197.6	216	*420*	788	443	*830*	1526
-11.7	*11*	51.8	11.1	*52*	125.6	33.9	*93*	199.4	221	*430*	806	449	*840*	1544
-11.1	*12*	53.6	11.7	*53*	127.4	34.4	*94*	201.2	227	*440*	824	454	*850*	1562
-10.6	*13*	55.4	12.2	*54*	129.2	35.0	*95*	203.0	232	*450*	842	460	*860*	1580
-10.0	*14*	57.2	12.8	*55*	131.0	35.6	*96*	204.8	238	*460*	860	466	*870*	1598
-9.4	*15*	59.0	13.3	*56*	132.8	36.1	*97*	206.6	243	*470*	878	471	*880*	1616
-8.9	*16*	60.8	13.9	*57*	134.6	36.7	*98*	208.4	249	*480*	896	477	*890*	1634
-8.3	*17*	62.6	14.4	*58*	136.4	37.2	*99*	210.2	254	*490*	914	482	*900*	1652
-7.8	*18*	64.4	15.0	*59*	138.2	37.8	*100*	212.0	260	*500*	932	488	*910*	1670
-7.2	*19*	66.2	15.6	*60*	140.0	43	*110*	230	266	*510*	950	493	*920*	1688
-6.7	*20*	68.0	16.1	*61*	141.8	49	*120*	248	271	*520*	968	499	*930*	1706
-6.1	*21*	69.8	16.7	*62*	143.6	54	*130*	266	277	*530*	986	504	*940*	1724
-5.6	*22*	71.6	17.2	*63*	145.4	60	*140*	284	282	*540*	1004	510	*950*	1742
-5.0	*23*	73.4	17.8	*64*	147.2	66	*150*	302	288	*550*	1022	516	*960*	1760
-4.4	*24*	75.2	18.3	*65*	149.0	71	*160*	320	293	*560*	1040	521	*970*	1778
-3.9	*25*	77.0	18.9	*66*	150.8	77	*170*	338	299	*570*	1058	527	*980*	1796
-3.3	*26*	78.8	19.4	*67*	152.6	82	*180*	356	304	*580*	1076	532	*990*	1814
-2.8	*27*	80.6	20.0	*68*	154.4	88	*190*	374	310	*590*	1094	538	*1000*	1832
-2.2	*28*	82.4	20.6	*69*	156.2	93	*200*	392	316	*600*	1112			
-1.7	*29*	84.2	21.1	*70*	158.0	99	*210*	410	321	*610*	1130			
-1.1	*30*	86.0	21.7	*71*	159.8	100	*212*	414	327	*620*	1148			

(Courtesy Bodine)

Electrical Formulas

Ohm's Law: Amperes $= \dfrac{\text{Volts}}{\text{Ohms}}$

Power in DC Circuits:

Watts $=$ Volts $\times$ Amperes

Horsepower $= \dfrac{\text{Volts} \times \text{Amperes}}{746}$

Kilowatts $= \dfrac{\text{Volts} \times \text{Amperes}}{1000}$

Kilowatt-Hours (kWh) $= \dfrac{\text{Volts} \times \text{Amperes} \times \text{Hours}}{1000}$

Power in AC Circuits:

Apparent power: Kilovolt-Amperes (kVa) $= \dfrac{\text{Volts} \times \text{Amperes}}{1000}$

Power Factor $= \dfrac{\text{Kilowatts}}{\text{Kilovolts-Amperes}}$

Single-Phase Kilowatts (kW) $=$

$$\dfrac{\text{Volts} \times \text{Amperes} \times \text{Power Factor}}{1000}$$

Two-Phase (kW) $=$
$$\dfrac{\text{Volts} \times \text{Amperes} \times \text{Power Factor} \times 1.4142}{1000}$$

Three-Phase (kW) $=$
$$\dfrac{\text{Volts} \times \text{Amperes} \times \text{Power Factor} \times 1.7321}{1000}$$

SI (Metric) Conversion Table

	SI Unit	Imperial/Metric to SI	SI to Imperial/Metric
Length	meter (m)	1 inch = 2.54×10^{-2} m 1 foot = 0.305 m 1 yard = .914 m	1 m = 39.37 inches = 3.281 feet = 1.094 yards
Mass	kilogram (kg.)	1 ounce (mass) = 28.35×10^{-3} kg. 1 pound (mass) = 0.454 kg. 1 slug = 14.59 kg.	1 kg. = 35.27 ounces = 2.205 pounds = 68.521×10^{-3} slug
Area	square meter (m^2)	1 sq. in. = 6.45×10^{-4} m^2 1 sq. ft. = 0.93×10^{-1} m^2 1. sq. yd. = 0.836 m^2	1 m^2 = 1550 sq. in. = 10.76 sq. ft. = 1.196 sq. yd.
Volume	cubic meter (m^3)	1 cu. in. = 16.3×10^{-6} m^3 1 cu. ft. = 0.028 m^3	1 m^3 = 6.102×10^4 cu. in. = 35.3 cu. ft.
Time	second (s)	same as Imperial/Metric	same as Imperial/Metric
Electric Current	ampere (A)	same as Imperial/Metric	same as Imperial/Metric
Plane Angle	radian (rad.)	1 angular deg. = 1.745×10^{-2} rad. 1 revolution = 6.283 rad.	1 r. = 57.296 rad.
Frequency	hertz (Hz.)	1 cycle/sec. = 1 Hz.	1 Hz. = 1 cps
Force	newton (N)	1 oz. (f) = 0.278 N 1 lb. (f) = 4.448 N 1 kilopond = 9.807 N 1 kgf = 9.807 N	1 N = 3.597 oz. (f) = 0.225 lb. (f) = 0.102 kp = 0.102 kgf
Energy (Work)	joule (J)	1 Btu = 1055.06 J 1 kWh = 3.6×10^6 J 1 Ws = 1 J 1 kcal = 4186.8 J	1 J = 9.478×10^{-4} Btu = 2.778×10^{-7} kWh = 1 Ws = 2.389×10^{-4} kcal
Power	watt (W)	1 hp (electric) = 746 W	1 W = 1.341×10^{-3} hp (electric)
Quantity of Electricity	coulomb (C)	same as Imperial/Metric	same as Imperial/Metric
EMF	volt (V)	same as Imperial/Metric	same as Imperial/Metric
Resistance	ohm (Ω)	same as Imperial/Metric	same as Imperial/Metric
Electric Capacitance	farad (F)	same as Imperial/Metric	same as Imperial/Metric
Electric Induction	henry (H)	same as Imperial/Metric	same as Imperial/Metric
Magnetic Flux	weber (Wb)	1 line = 10^{-8} Wb 1 Mx = 10^{-8} Wb 1 Vs = 1 Wb	1 Wb = 10^8 lines = 10^8 Mx = 1 Vs
Magnetic Flux Density	tesla (T)	1 line/$in.^2$ = 1.55×10^{-5} T 1 gauss = 10^{-4} T	1 T = 6.452×10^4 lines/$in.^2$ = 10^4 gauss
Linear Velocity	meter/sec. (m/s)	1 inch/sec. = 2.54×10^{-2} m/s 1 mph = 1.609 km/s	1 m/s = 39.37 in./sec. = 3.281 ft./sec.
Linear Accel.	meter/$sec.^2$ (m/s^2)	1 in./$sec.^2$ = 2.54×10^{-4} m/s^2	1 m/s^2 = 39.37 in./$sec.^2$
Torque	newtonmeter (N·m)	1 lb. ft. = 1.356 N·m 1 oz. in. = 7.062×10^{-3} N·m 1 kilopondmeter = 9.807 N·m 1 lb. in. = 0.113 N·m	1 N·m = 0.738 lb. ft. = 8.851 lb. in. = 0.102 kpm = 141.61 oz. in.
Temperature	degree Celsius (°C)	$F = (C \times \frac{9}{5}) + 32$	$C = (F - 32) \times \frac{5}{9}$

NOTE: There is at press time no international equivalent for Revolutions Per Minute (RPM).
Commonly used expressions are: RPM = r/min = t/min = U/min = Rev/min = min^{-1}
(f) = force

Metric Conversion Table

Fractions	Inches	mm	Fractions	Inches	mm
1/64	.0156	.3969	33/64	.5156	13.097
1/32	.0312	.7937	17/32	.5312	13.494
3/64	.0468	1.191	35/64	.5468	13.891
1/16	.0625	1.588	9/16	.5625	14.288
5/64	.0781	1.984	37/64	.5781	14.684
3/32	.0937	2.381	19/32	.5937	15.081
7/64	.1093	2.778	39/64	.6093	15.478
1/8	.125	3.175	5/8	.625	15.875
9/64	.1406	3.572	41/64	.6406	16.272
5/32	.1562	3.969	21/32	.6562	16.669
11/64	.1718	4.366	43/64	.6718	17.066
3/16	.1875	4.763	11/16	.6875	17.463
13/64	.2031	5.159	45/64	.7031	17.859
7/32	.2187	5.556	23/32	.7187	18.256
15/64	.2343	5.953	47/64	.7343	18.653
1/4	.25	6.350	3/4	.75	19.050
17/64	.2656	6.747	49/64	.7656	19.447
9/32	.2812	7.144	25/32	.7812	19.844
19/64	.2968	7.541	51/64	.7969	20.241
5/16	.3125	7.938	13/16	.8125	20.638
21/64	.3281	8.334	53/64	.8281	21.034
11/32	.3437	8.731	27/32	.8437	21.431
23/64	.3593	9.128	55/64	.8593	21.828
3/8	.375	9.525	7/8	.875	22.225
25/64	.3906	9.922	57/64	.8906	22.622
13/32	.4062	10.319	29/32	.9062	23.019
27/64	.4219	10.716	59/64	.9218	23.416
7/16	.4375	11.113	15/16	.9375	23.813
29/64	.4531	11.509	61/64	.9531	24.209
15/32	.4687	11.906	31/32	.9687	24.606
31/64	.4843	12.303	63/64	.9843	25.003
1/2	.5	12.700			

1 mm = .03937 inches
1 inch = 25.4 mm
1 meter = 3.2809 feet

Mechanical and Electrical Characteristics

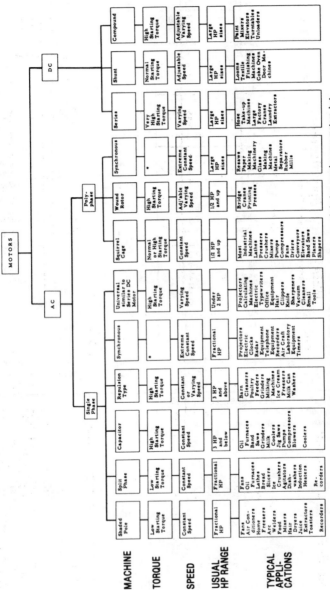

MOTORS

AC

Single Phase

	Shaded Pole	Split Phase	Capacitor	Repulsion Type	Synchronous	Universal similar to Series DC Motor
MACHINE	Shaded Pole	Split Phase	Capacitor	Repulsion Type	Synchronous	Universal similar to Series DC Motor
TORQUE	Low Starting Torque	Low Starting Torque	High Starting Torque	High Starting Torque	*	High Starting Torque
SPEED	Constant Speed	Constant Speed	Constant Speed	Constant varying Speed	Extreme Constant Speed	Varying Speed
USUAL HP RANGE	Fractional HP	Fractional HP	3 HP and below	3 HP and above	Fractional HP	Under 2 HP
TYPICAL APPLICATIONS	Fans, Air Conditioners, Home Freezers, Food Mixers, Arc Welders, Hair Dryers, Juice Extractors, Toasters, Recorders	Fans, Oil Furnaces, Lathes, Bread Slicers, Ice Crushers, Agitators, Dish-washers, Induction Heaters, Re-corders	Oil Furnaces, Band Saws, Grinders, Milk Coolers, Jig Saws, Pumps, Compressors, Blowers, Coolers	Barn Cleaners, Poultry Feeders, Grinders, Milking Machines, Ice Cream Freezers, Milk Can Washers	Projectors, Electric Clocks, Radar Equipment, Telephone Equipment, Recorders, Air Craft, Laboratory Equipment, Timers	Projectors, Calculating Machines, Electric Typewriters, Office Equipment, Hair Clippers, Knife Sharpeners, Vacuum Cleaners, Small Tools

Poly-phase

	Squirrel Cage	Wound Rotor	Synchronous
MACHINE	Squirrel Cage	Wound Rotor	Synchronous
TORQUE	Normal or High Starting Torque	High Starting Torque	*
SPEED	Constant Speed	Adjustable Varying Speed	Extreme Constant Speed
USUAL HP RANGE	1/2 HP and up	1/2 HP and up	Large HP sizes
TYPICAL APPLICATIONS	Most Industrial Machines, Lathes, Presses, Crushers, Mixers, Pumps, Compressors, Fans, Conveyors, Elevators, Band Saws, Planers, Shapers	Bridge Cranes, Printing Presses	Reswa Take-up Making Machinery, Glass Making Machine, Metal Separators, Rubber Mills

DC

	Series	Shunt	Compound
MACHINE	Series	Shunt	Compound
TORQUE	Very High Starting Torque	Normal Starting Torque	High Starting Torque
SPEED	Varying Speed	Adjustable Speed	Adjustable Varying Speed
USUAL HP RANGE	Large HP sizes	Large HP sizes	Large HP sizes
TYPICAL APPLICATIONS	Hoists, Take-up Machines, Large Factory Cranes, Laundry Extractors	Looms, Textile Finishing, Coke Oven Door Machines	Paint Mixers, Elevators, Turntables, Unloaders

* Starting Torque depends on starting method. Synchronous motors cannot start as synchronous motors, but must be started as one of the other types of AC motors.

(Courtesy Lincoln Electric)

Motor Frame Dimensions

NEMA Frame Specifications

Courtesy Leeson Electric Corp.

NEMA 48 Frame Rigid Base

OPEN DRIP-PROOF

FRAME	G48	J48	L48	N48	R48
c	8½	9	9½	10	10½
ad	2⅜	2⅞	3⅜	3⅞	4⅜

T.E.N.V.

FRAME	J48	L48	N48	R48	U48
c	9⅞	10⅜	10⅞	11⅜	11⅞
ad	3¼	3¾	4¼	4¾	5¼

NEMA 48 Frame Resilient Base

OPEN DRIP-PROOF

FRAME	G48	J48	L48	R48
c	9⅜	9⅞	10⅜	11⅜
ad	2¾	3¼	3¾	4¾

For totally enclosed non-ventilated designs use open drip-proof dimensions.

DRWG. NO.

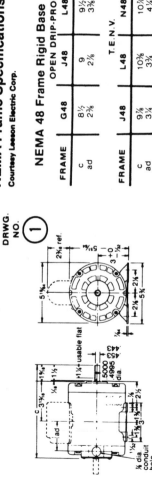

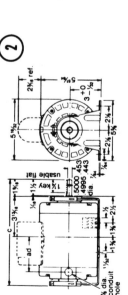

Motor Frame Dimensions (cont.)

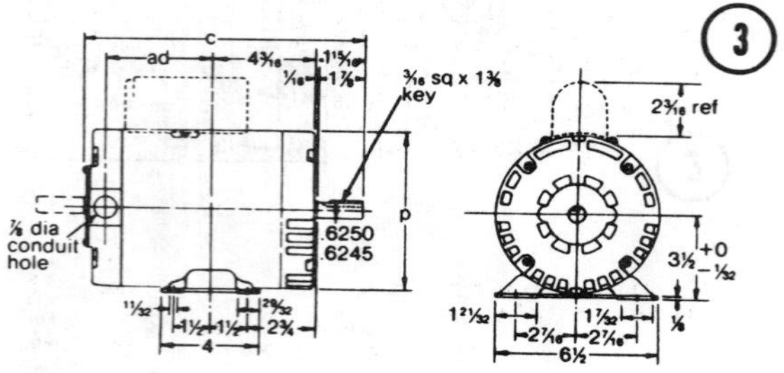

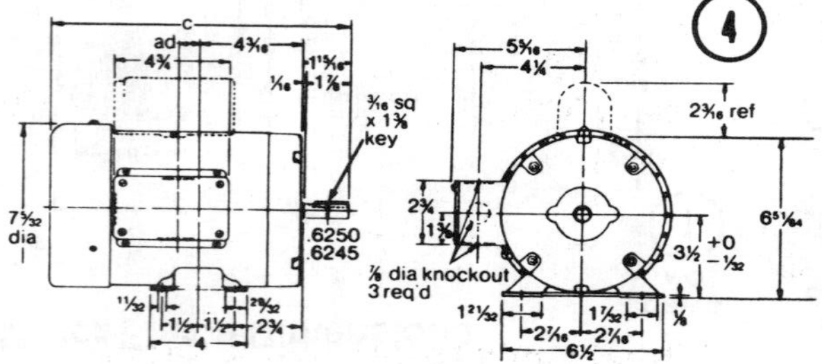

NEMA Frame Specifications (cont.)

NEMA 56 Frame Rigid Base

OPEN DRIP-PROOF

FRAME	JS56	LS56	NS56	RS56	C56	D56
c	9 1/2	10	10 1/2	11	9 7/8	10 3/8
ad	2 7/16	2 15/16	3 7/16	3 15/16	2 7/8	3 3/8
p	5 19/32	5 19/32	5 19/32	5 19/32	6 19/32	6 19/32
FRAME	**E56**	**F56**	**G56**	**H56**	**J56**	**K56**
c	10 7/8	11 3/8	11 7/8	12 3/8	12 7/8	13 3/8
ad	3 7/8	4 3/8	4 7/8	5 3/8	5 7/8	6 3/8
p	6 19/32	6 19/32	6 19/32	6 19/32	6 19/32	6 19/32

For totally enclosed non-ventilated designs use open drip-proof dimensions. For S56 totally enclosed non-ventilated designs only, add 1" to "c" dimension.

T.E.F.C.

FRAME	C56	D56	E56	F56	G56	H56	J56	K56
c	10 13/16	11 5/16	11 13/16	12 5/16	12 13/16	13 5/16	13 13/16	14 5/16
ad	−9/16	−1/16	7/16	15/16	1 7/16	1 15/16	2 7/16	2 15/16

Motor Frame Dimensions (cont.)

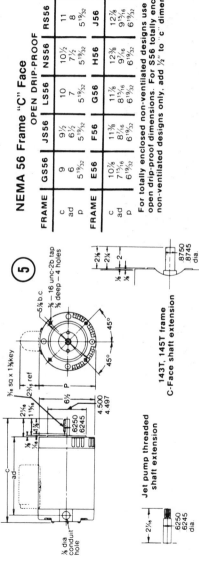

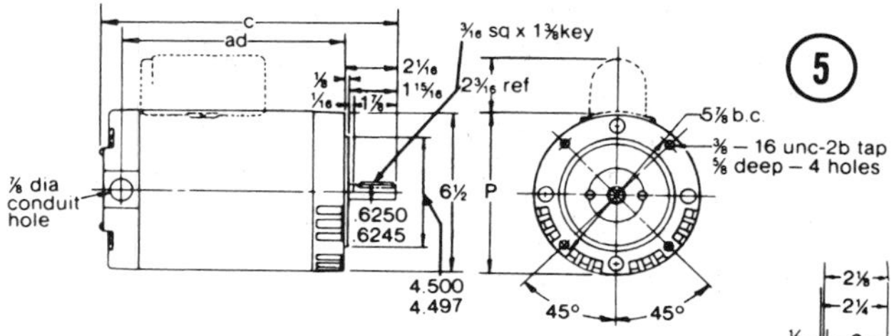
5
c
ad
⅜ dia conduit hole
3/16 sq x 1⅜ key
2 1/16
1 15/16
⅛
1/16
1⅞
2 3/16 ref
6½
P
.6250
.6245
4.500
4.497
5⅞ b.c.
⅜ – 16 unc-2b tap
⅝ deep – 4 holes
45°
45°

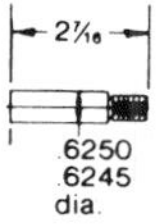
2 7/16
.6250
.6245
dia.

Jet pump threaded shaft extension

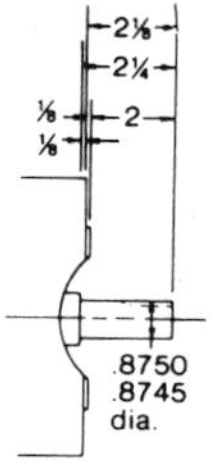
2⅛
2¼
⅛
⅛
2
.8750
.8745
dia.

143T, 145T frame C-Face shaft extension

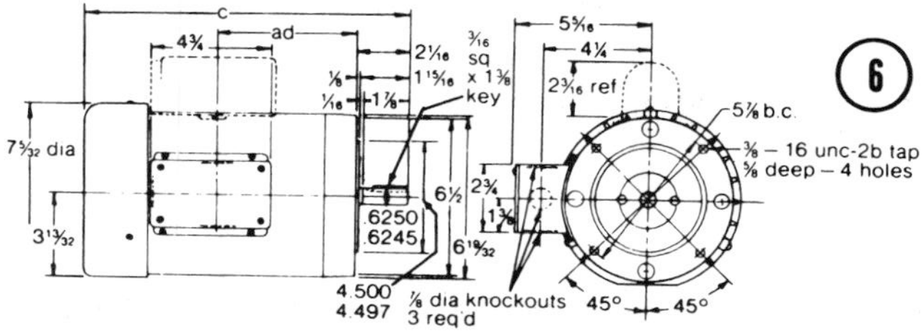
6
c
ad
4¾
7 5/32 dia
3 13/32
2 1/16
1 15/16
⅛
1/16
1⅞
3/16 sq x 1⅜ key
.6250
.6245
6½
2¾
1⅜
6 19/32
4.500
4.497
⅞ dia knockouts 3 req'd
5 5/16
4¼
2 3/16 ref
5⅞ b.c.
⅜ – 16 unc-2b tap
⅝ deep – 4 holes
45°
45°

NEMA Frame Specifications (cont.)

NEMA 56 Frame "C" Face

OPEN DRIP-PROOF

FRAME	GS56	JS56	LS56	NS56	RS56	C56	D56
c	9	9½	10	10½	11	9⅞	10⅜
ad	6	6½	7	7½	8	6 15/16	7 7/16
p	5 19/32	5 19/32	5 19/32	5 19/32	5 19/32	6 19/32	6 19/32

FRAME	E56	F56	G56	H56	J56	K56
c	10⅞	11⅜	11⅞	12⅜	12⅞	13⅜
ad	7 15/16	8 7/16	8 15/16	9 7/16	9 15/16	10 7/16
p	6 19/32	6 19/32	6 19/32	6 19/32	6 19/32	6 19/32

For totally enclosed non-ventilated designs use open drip-proof dimensions. For S56 totally enclosed non-ventilated designs only, add ½" to "c" dimension.

T. E. F. C.

FRAME	C56	D56	E56	F56	G56	H56	J56	K56
c	10 13/16	11 5/16	11 13/16	12 5/16	12 13/16	13 5/16	13 13/16	14 5/16
ad	3½	4	4½	5	5½	6	6½	7

Motor Frame Dimensions (cont.)

NEMA Frame Specifications (cont.)

NEMA 56 Frame Resilient Base

OPEN DRIP-PROOF

FRAME	GS56	JS56	LS56	NS56	RS56	C56
c	$9\frac{13}{16}$	$10\frac{5}{8}$	$10\frac{13}{16}$	$11\frac{5}{8}$	$11\frac{13}{16}$	$10\frac{13}{16}$
ad	$2\frac{7}{8}$	$2\frac{15}{16}$	$3\frac{5}{8}$	$3\frac{15}{16}$	$4\frac{1}{8}$	$3\frac{15}{32}$
p	$5\frac{19}{32}$	$5\frac{19}{32}$	$5\frac{19}{32}$	$5\frac{19}{32}$	$5\frac{19}{32}$	$6\frac{19}{32}$

FRAME	D56	E56	F56	G56	J56
c	$11\frac{5}{16}$	$11\frac{13}{16}$	$12\frac{5}{8}$	$12\frac{13}{16}$	$13\frac{13}{16}$
ad	$3\frac{31}{32}$	$4\frac{15}{32}$	$4\frac{3}{32}$	$5\frac{15}{32}$	$6\frac{15}{32}$
p	$6\frac{19}{32}$	$6\frac{19}{32}$	$6\frac{19}{32}$	$6\frac{19}{32}$	$6\frac{19}{32}$

For totally enclosed non-ventilated designs use open drip-proof dimensions.

NEMA Combination Base 56HZ, 143T, 145T

OPEN DRIP-PROOF

FRAME	E143T	F143T	G145T	H145T	J145T	K145T	L145T
c	$11\frac{1}{16}$	$11\frac{9}{16}$	$12\frac{1}{8}$	$12\frac{9}{16}$	$13\frac{1}{8}$	$13\frac{9}{16}$	$14\frac{1}{16}$
ad	$2\frac{3}{8}$	$2\frac{11}{16}$	$3\frac{1}{16}$	$3\frac{9}{16}$	$4\frac{1}{16}$	$4\frac{9}{16}$	$5\frac{1}{16}$

(7)

(8)

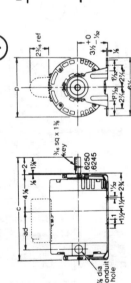

Motor Frame Dimensions (cont.)

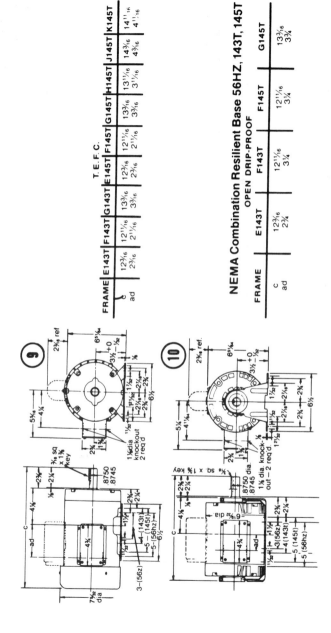

NEMA Frame Specifications (cont.)

T.E.F.C.

FRAME	E143T	F143T	G143T	E145T	F145T	G145T	H145T	J145T	K145T
c	12 3/16	12 11/16	13 3/16	12 3/16	12 11/16	13 3/16	13 11/16	14 3/16	14 11/16
ad	2 3/16	2 11/16	3 3/16	2 3/16	2 11/16	3 3/16	3 11/16	4 3/16	4 11/16

NEMA Combination Resilient Base 56HZ, 143T, 145T
OPEN DRIP-PROOF

FRAME	E143T	F143T	G143T	F145T	G145T
c	12 3/16	12 11/16	13 3/16	12 11/16	13 3/16
ad	2 3/4	3 1/4	3 3/4	3 1/4	3 3/4

NEMA Frame Specifications (cont.)

Motor Frame Dimensions (cont.)

NEMA 56 Frame Rigid Base Brakemotor

OPEN DRIP-PROOF

FRAME	C56	D56	E56	F56	G56
c	13⅞₆	14¹⁄₁₆	14⁹⁄₁₆	15¹⁄₁₆	15⁹⁄₁₆
ad	−⁹⁄₁₆	−¹⁄₁₆	⁷⁄₁₆	¹⁵⁄₁₆	1⁷⁄₁₆

For totally enclosed non-ventilated designs use
open drip-proof dimensions

NEMA 56 Frame "C" Face Brakemotor

OPEN DRIP-PROOF

FRAME	C56	D56	E56	F56	G56
c	13⅞₆	14¹⁄₁₆	14⁹⁄₁₆	15¹⁄₁₆	15⁹⁄₁₆
ad	−⁹⁄₁₆	−¹⁄₁₆	⁷⁄₁₆	¹⁵⁄₁₆	1⁷⁄₁₆

For totally enclosed non-ventilated designs use
open drip-proof dimensions

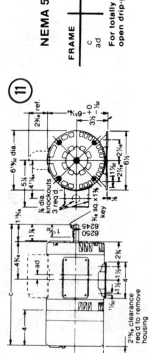

Motor Frame Dimensions (cont.)

Threaded shaft extension

NEMA Frame Specifications (cont.)

(13)

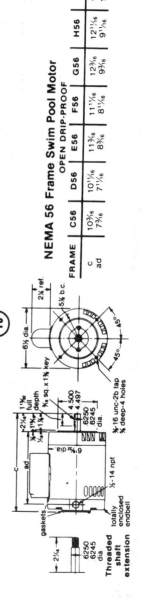

NEMA 56 Frame Swim Pool Motor

OPEN DRIP-PROOF

FRAME	C56	D56	E56	F56	G56	H56	K56
c	10³⁄₁₆	10¹¹⁄₁₆	11³⁄₁₆	11¹¹⁄₁₆	12³⁄₁₆	12¹¹⁄₁₆	13¹¹⁄₁₆
ad	7³⁄₁₆	7¹¹⁄₁₆	8³⁄₁₆	8¹¹⁄₁₆	9³⁄₁₆	9¹¹⁄₁₆	10¹¹⁄₁₆

Electronics Symbols
for
Devices Used with Electric Motors

NAME	SYMBOL
DIODE	Anode ▷|− Cathode
ZENER DIODE	
TRANSISTOR	B — C / E
SCR (SILICON CONTROLLED RECTIFIER)	Anode / Gate / Cathode

NAME	SYMBOL

BATTERY

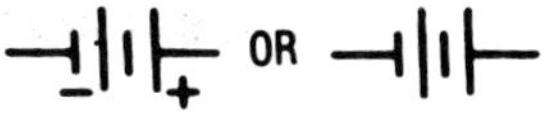

NEW   OLD

CAPACITOR FIXED

CIRCUIT BREAKERS

    Air Circuit Breaker

    Three-pole Power Circuit Breaker
    (Single-throw) (with Terminals)

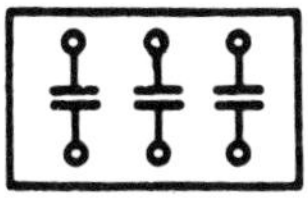

    Thermal Trip Air Circuit Breaker

COILS

    Nonmagnetic Core-fixed

    Magnetic Core-fixed

    Magnetic Core-adjustable Tap or
    Slide Wire

    Operating Coil

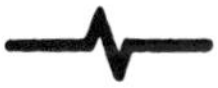

    Blowout Coil

    Blowout Coil with Terminals

    Series Field

    Shunt Field

    Commutating Field

NAME	SYMBOL

CONNECTIONS (MECHANICAL)

Mechanical Connection of Shield

Mechanical Interlock

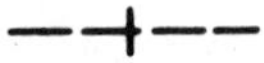

Direct Connected Units

CONNECTIONS (WIRING)

Electric Conductor—Control

Electric Conductor—Power

Junction of Conductors

Wiring Terminal

Ground

Crossing of Conductors—Not Connected

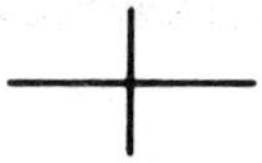

Crossing of Connected Conductors

Joining of Conductors—Not Crossing

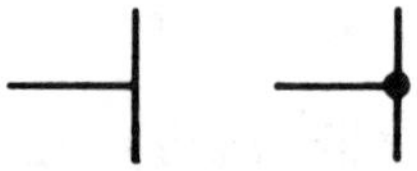

CONTACTS (ELECTRICAL)

Normally Closed Contact (NC)

Normally Open Contact (NO)

NO Contact With Time Closing (TC) Feature

NC Contact With Time Opening (TO) Feature

Note: NO (Normally Open) and NC (Normally Closed) Designates the Position of the Contacts When the Main Device is In the Deenergized or Nonoperated Position

NAME	SYMBOL

CONTACTOR, SINGLE-POLE ELECTRICALLY OPERATED, WITH BLOWOUT COIL

> Note: Fundamental Symbols for Contact, Coils, Mechanical Connections, etc., are the Basis of Contactor Symbols

FUSE

INDICATING LIGHTS

    Indicating Lamp with Leads

    Indicating Lamp with Terminals

INSTRUMENTS

    Ammeter, with Terminals OR

    Voltmeter, with Terminals OR

    Wattmeter, with Terminals OR

MACHINES (ROTATING)

    Machine or Rotating Armature

    Squirrel-Cage Induction Motor

    Wound-Rotor Induction Motor or Generator

    Synchronous Motor, Generator or Condenser

NAME	SYMBOL

DC Compound Motor or Generator

Note: Commutating, Series, and Shunt
Fields May be Indicated by
1, 2 and 3 Zigzags Respectively.
Series and Shunt Coils May be
Indicated by Heavy and Light
Lines or 1 and 2 Zigzags Re-
spectively.

WINDING SYMBOLS

Three-Phase Wye (Ungrounded)

Three-Phase Wye (Grounded)

Three-Phase Delta

Note: Winding Symbols May be Shown
in Circles for all Motor and
Generator Symbols.

RECTIFIER, DRY OR ELECTROLYTIC, FULL WAVE

RELAYS

Overcurrent or Overvoltage Relay
with 1 No Contact

Thermal Overload Relay Having
2 Series Heating Elements and
1 NC Contact

RESISTORS (OLD SYMBOLS)

Resistor, Fixed, with Leads

Resistor, Fixed, with Terminals

Resistor, Adjustable Tap or
Slide Wire

NAME	SYMBOL
Resistor, Adjustable by Fixed Leads	
Resistor, Adjustable by Fixed Terminals	
Instrument or Relay Shunt	

RESISTORS (New Symbols)

Resistor (Fixed)
Resistor (Variable)

SWITCHES

Knife Switch, Single-Pole (SP)	
Knife Switch, Double-Pole Single-Throw (DPST)	
Knife Switch, Triple-Pole Single-Throw (TPST)	
Knife Switch, Single-Pole Double-Throw (SPDT)	
Knife Switch, Double-Pole Double-Throw (DPDT)	
Knife Switch, Triple-Pole Double-Throw (TPDT)	
Field-Discharge Switch with Resistor	

Push Button Normally Open (NO)

Push Button Normally Closed (NC)

Push Button Open and Closed
(Spring-Return)

Normally Closed Limit Switch
Contact

Normally Open Limit Switch Contact

Thermal Element
(Fuse)

TRANSFORMERS

	(OLD)	(NEW)
Single-Phase, Two-Winding Transformer		
Autotransformer Single-Phase		

Index

**Over a Century of Excellence
for the Professional
and
Vocational Trades and the Crafts**

Order now from your local bookstore
or use the convenient order form
at the back of this book.

AUDEL

These fully illustrated, up-to-date guides and manuals mean a better job done for mechanics, engineers, electricians, plumbers, carpenters, and all skilled workers.

CONTENTS

ELECTRICAL

House Wiring (Seventh Edition)

ROLAND E. PALMQUIST;
revised by PAUL ROSENBERG

5 1/2 x 8 1/4   Hardcover   248 pp.   150 Illus.
ISBN: 0-02-594692-7 $22.95

Rules and regulations of the current 1990 National Electrical Code for residential wiring fully explained and illustrated.

Practical Electricity
(Fifth Edition)

ROBERT G. MIDDLETON;
revised by L. DONALD MEYERS

5 1/2 x 8 1/4   Hardcover   512 pp.   335 Illus.
ISBN: 0-02-584561-6 $19.95

The fundamentals of electricity for electrical workers, apprentices, and others requiring concise information about electric principles and their practical applications.

Guide to the 1990 National Electrical Code

ROLAND E. PALMQUIST;
revised by PAUL ROSENBERG

5 1/2 x 8 1/4   Hardcover   664 pp.   230 Illus.
ISBN: 0-02-594565-3 $24.95

The most authoritative guide available to interpreting the National Electrical Code for electricians, contractors, electrical inspectors, and homeowners. Examples and illustrations.

Installation Requirements of the 1990 National Electrical Code

PAUL ROSENBERG

5 1/2 × 8 1/4   Hardcover   240 pp.
ISBN: 0-02-604941-4 $24.95

Field guide for installation requirements makes understanding the 1990 Electrical Code simple while on the job. Applications and easy-to-understand tables make this the perfect working companion.

Mathematics for Electricians and Electronics Technicians

REX MILLER

5 1/2 x 8 1/4   Hardcover   312 pp.   115 Illus.
ISBN: 0-8161-1700-4 $14.95

Mathematical concepts, formulas, and problem-solving techniques utilized on-the-job by electricians and those in electronics and related fields.

Fractional-Horsepower Electric Motors

REX MILLER and
MARK RICHARD MILLER

5 1/2 x 8 1/4   Hardcover   436 pp.   285 Illus.
ISBN: 0-672-23410-6 $15.95

The installation, operation, maintenance, repair, and replacement of the small-to-moderate-size electric motors that power home appliances and industrial equipment.

Electric Motors (Fifth Edition)

EDWIN P. ANDERSON
and REX MILLER

5 1/2 x 8 1/4   Hardcover   696 pp.
Photos/line art
ISBN: 0-02-501920-1 $35.00

Complete reference guide for electricians, industrial maintenance personnel, and installers. Contains both theoretical and practical descriptions.

Home Appliance Servicing
(Fourth Edition)

EDWIN P. ANDERSON;
revised by REX MILLER

5 1/2 x 8 1/4   Hardcover   640 pp.   345 Illus.
ISBN: 0-672-23379-7 $22.50

The essentials of testing, maintaining, and repairing all types of home appliances.

Television Service Manual
(Fifth Edition)

ROBERT G. MIDDLETON;
revised by JOSEPH G. BARRILE

5 1/2 x 8 1/4   Hardcover   512 pp.   395 Illus.
ISBN: 0-672-23395-9 $16.95

A guide to all aspects of television transmission and reception, including the operating principles of black and white and color receivers. Step-by-step maintenance and repair procedures.

Electrical Course for Apprentices and Journeymen
(Third Edition)

ROLAND E. PALMQUIST

5 1/2 x 8 1/4   Hardcover   478 pp.   290 Illus.
ISBN: 0-02-594550-5 $19.95

This practical course in electricity for those in formal training programs or learning on their own provides a thorough understanding of operational theory and its applications on the job.

Questions and Answers for Electricians Examinations
(Tenth Edition)

Revised by PAUL ROSENBERG

5 1/2 x 8 1/4   Hardcover   316 pp.   110 Illus.
ISBN: 0-02-604955-4 $22.95

Based on the 1990 National Electrical Code, this book reviews the subjects included in the various electricians examinations—apprentice, journeyman, and master.

> # MACHINE SHOP AND MECHANICAL TRADES

Machinists Library
(Fourth Edition, 3 Vols.)

REX MILLER

5 1/2 x 8 1/4 Hardcover 1,352 pp. 1120 Illus.
ISBN: 0-672-23380-0 $52.95

An indispensable three-volume reference set for machinists, tool and die makers, machine operators, metal workers, and those with home workshops. The principles and methods of the entire field are covered in an up-to-date text, photographs, diagrams, and tables.

Volume I: Basic Machine Shop

REX MILLER

5 1/2 x 8 1/4   Hardcover   392 pp.   375 Illus.
ISBN: 0-672-23381-9 $17.95

Volume II: Machine Shop

REX MILLER

5 1/2 x 8 1/4   Hardcover   528 pp.   445 Illus.
ISBN: 0-672-23382-7 $19.95

Volume III: Toolmakers Handy Book

REX MILLER

5 1/2 x 8 1/4   Hardcover   432 pp.   300 Illus.
ISBN: 0-672-23383-5 $14.95

Mathematics for Mechanical Technicians and Technologists

JOHN D. BIES

5 1/2 x 8 1/4   Hardcover   342 pp.   190 Illus.
ISBN: 0-02-510620-1 $17.95

The mathematical concepts, formulas, and problem-solving techniques utilized on the job by engineers, technicians, and other workers in industrial and mechanical technology and related fields.

Millwrights and Mechanics Guide (Fourth Edition)

CARL A. NELSON

5 1/2 x 8 1/4   Hardcover   1,040 pp.   880 Illus.
ISBN: 0-02-588591-x $29.95

The most comprehensive and authoritative guide available for millwrights, mechanics, maintenance workers, riggers, shop workers, foremen, inspectors, and superintendents on plant installation, operation, and maintenance.

Welders Guide (Third Edition)
JAMES E. BRUMBAUGH

5 1/2 x 8 1/4   Hardcover   960 pp.   615 Illus.
ISBN: 0-672-23374-6 $23.95

The theory, operation, and maintenance of all welding machines. Covers gas welding equipment, supplies, and process; arc welding equipment, supplies, and process; TIG and MIG welding; and much more.

Welders/Fitters Guide
HARRY L. STEWART

8 1/2 x 11   Paperback   160 pp.   195 Illus.
ISBN: 0-672-23325-8 $7.95

Step-by-step instruction for those training to become welders/fitters who have some knowledge of welding and the ability to read blueprints.

Sheet Metal Work
JOHN D. BIES

5 1/2 x 8 1/4   Hardcover   456 pp.   215 Illus.
ISBN: 0-8161-1706-3 $19.95

An on-the-job guide for workers in the manufacturing and construction industries and for those with home workshops. All facets of sheet metal work detailed and illustrated by drawings, photographs, and tables.

Power Plant Engineers Guide
(Third Edition)
FRANK D. GRAHAM;
revised by CHARLIE BUFFINGTON

5 1/2 x 8 1/4   Hardcover   960 pp.   530 Illus.
ISBN: 0-672-23329-0 $27.50

This all-inclusive, one-volume guide is perfect for engineers, firemen, water tenders, oilers, operators of steam and diesel-power engines, and those applying for engineer's and firemen's licenses.

Mechanical Trades Pocket
Manual (Third Edition)
CARL A. NELSON

4 x 6   Paperback   364 pp.   255 Illus.
ISBN: 0-02-588665-7 $14.95

A handbook for workers in the industrial and mechanical trades on methods, tools, equipment, and procedures. Pocket-sized for easy reference and fully illustrated.

PLUMBING

Plumbers and Pipe Fitters
Library (Fourth Edition, 3 Vols.)
CHARLES N. McCONNELL

5 1/2 x 8 1/4   Hardcover   952 pp.   560 Illus.
ISBN: 0-02-582914-9 $68.45

This comprehensive three-volume set contains the most up-to-date information available for master plumbers, journeymen, apprentices, engineers, and those in the building trades. A detailed text and clear diagrams, photographs, and charts and tables treat all aspects of the plumbing, heating, and air conditioning trades.

Volume I: Materials, Tools, Roughing-In
CHARLES N. McCONNELL;
revised by TOM PHILBIN

5 1/2 x 8 1/4   Hardcover   304 pp.   240 Illus.
ISBN: 0-02-582911-4 $20.95

**Volume II: Welding, Heating,
Air Conditioning**
CHARLES N. McCONNELL;
revised by TOM PHILBIN

5 1/2 x 8 1/4   Hardcover   384 pp.   220 Illus.
ISBN: 0-02-582912-2 $22.95

**Volume III: Water Supply, Drainage,
Calculations**
CHARLES N. McCONNELL;
revised by TOM PHILBIN

5 1/2 x 8 1/4   Hardcover   264 pp.   100 Illus.
ISBN: 0-02-582913-0 $20.95

Home Plumbing Handbook
(Third Edition)
CHARLES N. McCONNELL

8 1/2 x 11   Paperback   200 pp.   100 Illus.
ISBN: 0-672-23413-0 $14.95

An up-to-date guide to home plumbing installation and repair.

The Plumbers Handbook
(Eighth Edition)
JOSEPH P. ALMOND, SR.;
revised by REX MILLER

4 × 6   Paperback   368 pp.   170 Illus
ISBN: 0-02-501570-2 $19.95

Comprehensive and handy guide for plumbers and pipefitters—fits in the toolbox or pocket. For apprentices, journeymen, or experts.

Questions and Answers for Plumbers' Examinations
(Third Edition)

JULES ORAVETZ;
revised by REX MILLER

*5 1/2 x 8 1/4 Paperback 288 pp. 145 Illus.
ISBN: 0-02-593510-0 $14.95*

Complete guide to preparation for the plumbers' exams given by local licensing authorities. Includes requirements of the National Bureau of Standards.

HVAC

Air Conditioning: Home and Commercial (Fourth Edition)

EDWIN P. ANDERSON;
revised by REX MILLER

*5 1/2 x 8 1/4 Hardcover 528 pp. 180 Illus.
ISBN: 0-02-584885-2 $29.95*

A guide to the construction, installation, operation, maintenance, and repair of home, commercial, and industrial air conditioning systems.

Heating, Ventilating, and Air Conditioning Library
(Second Edition, 3 Vols.)

JAMES E. BRUMBAUGH

*5 1/2 x 8 1/4 Hardcover 1,840 pp. 1,275 Illus.
ISBN: 0-672-23388-6 $53.85*

An authoritative three-volume reference library for those who install, operate, maintain, and repair HVAC equipment commercially, industrially, or at home.

Volume I: Heating Fundamentals, Furnaces, Boilers, Boiler Conversions

JAMES E. BRUMBAUGH

*5 1/2 x 8 1/4 Hardcover 656 pp. 405 Illus.
ISBN: 0-672-23389-4 $17.95*

Volume II: Oil, Gas and Coal Burners, Controls, Ducts, Piping, Valves

JAMES E. BRUMBAUGH

*5 1/2 x 8 1/4 Hardcover 592 pp. 455 Illus.
ISBN: 0-672-23390-8 $17.95*

Volume III: Radiant Heating, Water Heaters, Ventilation, Air Conditioning, Heat Pumps, Air Cleaners

JAMES E. BRUMBAUGH

*5 1/2 x 8 1/4 Hardcover 592 pp. 415 Illus.
ISBN: 0-672-23391-6 $17.95*

Oil Burners (Fifth Edition)

EDWIN M. FIELD

*5 1/2 x 8 1/4 Hardcover 360 pp. 170 Illus.
ISBN: 0-02-537745-0 $29.95*

An up-to-date sourcebook on the construction, installation, operation, testing, servicing, and repair of all types of oil burners, both industrial and domestic.

Refrigeration: Home and Commercial (Fourth Edition)

EDWIN P. ANDERSON;
revised by REX MILLER

*5 1/2 x 8 1/4 Hardcover 768 pp. 285 Illus.
ISBN: 0-02-584875-5 $34.95*

A reference for technicians, plant engineers, and the homeowner on the installation, operation, servicing, and repair of everything from single refrigeration units to commercial and industrial systems.

PNEUMATICS AND HYDRAULICS

Hydraulics for Off-the-Road Equipment (Second Edition)

HARRY L. STEWART;
revised by TOM PHILBIN

*5 1/2 x 8 1/4 Hardcover 256 pp. 175 Illus.
ISBN: 0-8161-1701-2 $13.95*

This complete reference manual on heavy equipment covers hydraulic pumps, accumulators, and motors; force components; hydraulic control components; filters and filtration, lines and fittings, and fluids; hydrostatic transmissions; maintenance; and troubleshooting.

Pneumatics and Hydraulics
(Fourth Edition)

HARRY L. STEWART;
revised by TOM STEWART

*5 1/2 x 8 1/4 Hardcover 512 pp. 315 Illus.
ISBN: 0-672-23412-2 $19.95*

The principles and applications of fluid power. Covers pressure, work, and power; general features of machines; hydraulic and pneumatic symbols; pressure boosters; air compressors and accessories; and much more.

Pumps (Fifth Edition)

HARRY L. STEWART;
revised by REX MILLER

5 1/2 x 8 1/4 Hardcover 552 pp. 360 Illus.
ISBN: 0-02-614725-4 $35.00

The practical guide to operating principles of pumps, controls, and hydraulics. Covers installation and day-to-day service.

CARPENTRY AND
CONSTRUCTION

Carpenters and Builders Library
(Sixth Edition, 4 Vols.)

JOHN E. BALL;
revised by JOHN LEEKE

5 1/2 x 8 1/4 Hardcover 1,300 pp. 988 Illus.
ISBN: 0-02-506455-4 $89.95

This comprehensive four-volume library has set the professional standard for decades for carpenters, joiners, and woodworkers.

Volume 1: Tools, Steel Square, Joinery

JOHN E. BALL;
revised by JOHN LEEKE

5 1/2 x 8 1/4 Hardcover 377 pp. 340 Illus.
ISBN: 0-02-506451-7 $21.95

Volume 2: Builders Math, Plans, Specifications

JOHN E. BALL;
revised by JOHN LEEKE

5 1/2 x 8 1/4 Hardcover 319 pp. 200 Illus.
ISBN: 0-02-506452-5 $21.95

Volume 3: Layouts, Foundation, Framing

JOHN E. BALL;
revised by JOHN LEEKE

5 1/2 x 8 1/4 Hardcover 269 pp. 204 Illus.
ISBN: 0-02-506453-3 $21.95

Volume 4: Millwork, Power Tools, Painting

JOHN E. BALL;
revised by JOHN LEEKE

5 1/2 x 8 1/4 Hardcover 335 pp. 244 Illus.
ISBN: 0-02-506454-1 $21.95

Complete Building Construction
(Second Edition)

JOHN PHELPS;
revised by TOM PHILBIN

5 1/2 x 8 1/4 Hardcover 744 pp. 645 Illus.
ISBN: 0-672-23377-0 $22.50

Constructing a frame or brick building from the footings to the ridge. Whether the building project is a tool shed, garage, or a complete home, this single fully illustrated volume provides all the necessary information.

Complete Roofing Handbook

JAMES E. BRUMBAUGH

5 1/2 x 8 1/4 Hardcov___ ___ Illus.
ISBN: 0-02-517____

NEW EDITION FOR 1992

Cov___ ___ ___ ___ofing; ro___ ___ ___ ntilation; sky- lig___ ___ ___ings; dormer construc- tio___ ___ ing details; and much more.

Complete Siding Handbook

JAMES E. BRUMBAUGH

5 1/2 x 8 1/4 Hardcover 512 pp. 450 Illus.
ISBN: 0-02-517880-6 $24.9_

This compa___ ___ ___ plete ___ ___ ___ nsive s___ ___ ___mpanying li___ ___y aspect of siding a b___

NEW EDITION FOR 1992

Masons and Builders Library
(Second Edition, 2 Vols.)

LOUIS M. DEZETTEL;
revised by TOM PHILBIN

5 1/2 x 8 1/4 Hardcover 688 pp. 500 Illus
ISBN: 0-672-23401-7 $27.95

This two-volume set provides practical instruction in bricklaying and masonry. Covers brick; mortar; tools; bonding; corners, openings, and arches; chimneys and fireplaces; structural clay tile and glass block; brick walls; and much more.

Volume 1: Concrete, Block, Tile, Terrazzo

LOUIS M. DEZETTEL;
revised by TOM PHILBIN

5 1/2 x 8 1/4 Hardcover 304 pp. 190 Illus
ISBN: 0-672-23402-5 $14.95

Volume 2: Bricklaying, Plastering, Rock Masonry, Clay Tile

LOUIS M. DEZETTEL;
revised by TOM PHILBIN

5 1/2 x 8 1/4 Hardcover 384 pp. 310 Illus
ISBN: 0-672-23403-3 $14.95

WOODWORKING

Wood Furniture: Finishing, Refinishing, Repairing
(Second Edition)

JAMES E. BRU___

NEW EDITION FOR 1992

A fully illustrated guide to repairing furniture and finishing and refinishing wood surfaces. Covers tools and supplies; types of wood; veneering; inlaying; repairing, restoring, and stripping; wood preparation; and much more.

Woodworking and Cabinetmaking
F. RICHARD BOLLER

5 1/2 x 8 1/4   Hardcover   360 pp.   455 Illus.
ISBN: 0-02-512800-0 $18.95

Essential information on all aspects of working with wood. Step-by-step procedures for woodworking projects are accompanied by detailed drawings and photographs.

MAINTENANCE AND REPAIR

Building Maintenance
(Second Edition)
JULES ORAVETZ

5 1/2 x 8 1/4   Paperback   384 pp.   210 Illus.
ISBN: 0-672-23278-2 $11.95

Professional maintenance procedures used in office, educational, and commercial buildings. Covers painting and decorating; plumbing and pipe fitting; concrete and masonry; and much more.

Gardening, Landscaping and Grounds Maintenance
(Third Edition)
JULES ORAVETZ

5 1/2 x 8 1/4   Hardcover   424 pp.   340 Illus.
ISBN: 0-672-23417-3 $15.95

Maintaining lawns and gardens as well as industrial, municipal, and estate grounds.

Home Maintenance and Repair: Walls, Ceilings and Floors
GARY D. BRANSON

8 1/2 x 11   Paperback   80 pp.   80 Illus.
ISBN: 0-672-23281-2 $6.95

The do-it-yourselfer's guide to interior remodeling with professional results.

Painting and Decorating
REX MILLER and GLEN E. BAKER

5 1/2 x 8 1/4   Hardcover   464 pp.   325 Illus.
ISBN: 0-672-23405-x $18.95

A practical guide for painters, decorators, and homeowners to the most up-to-date materials and techniques in the field.

Tree Care (Second Edition)
JOHN M. HALLER

8 1/2 x 11   Paperback   224 pp.   305 Illus.
ISBN: 0-02-062870-6 $16.95

The standard in the field. A comprehensive guide for growers, nursery owners, foresters, landscapers, and homeowners to planting, nurturing, and protecting trees.

Upholstering (Updated)
JAMES E. BRUMBAUGH

5 1/2 x 8 1/4 ... Illus.
ISBN: ...

NEW EDITION FOR 1992

... fully explained ... the professional, the approach ... and the hobbyist.

AUTOMOTIVE AND ENGINES

Diesel Engine Manual
(Fourth Edition)
PERRY O. BLACK;
revised by WILLIAM E. SCAHILL

5 1/2 x 8 1/4   Hardcover   512 pp.   255 Illus.
ISBN: 0-672-23371-1 $15.95

The principles, design, operation, and maintenance of today's diesel engines. All aspects of typical two- and four-cycle engines are thoroughly explained and illustrated by photographs, line drawings, and charts and tables.

Gas Engine Manual
(Third Edition)
EDWIN P. ANDERSON;
revised by CHARLES G. FACKLAM

5 1/2 x 8 1/4   Hardcover   424 pp.   225 Illus.
ISBN: 0-8161-1707-1 $12.95

How to operate, maintain, and repair gas engines of all types and sizes. All engine parts and step-by-step procedures are illustrated by photographs, diagrams, and troubleshooting charts.

Small Gasoline Engines
REX MILLER and
MARK RICHARD MILLER

5 1/2 x 8 1/4   Hardcover   640 pp.   525 Illus.
ISBN: 0-672-23414-9 $16.95

Practical information for those who repair, maintain, and overhaul two- and four-cycle engines—including lawn mowers, edgers, grass sweepers, snowblowers, emergency electrical generators, outboard motors, and other equipment with engines of up to ten horsepower.

Truck Guide Library (3 Vols.)
JAMES E. BRUMBAUGH

*5 1/2 x 8 1/4 Hardcover 2,144 pp. 1,715 Illus.
ISBN: 0-672-23392-4 $50.95*

This three-volume set provides the most comprehensive, profusely illustrated collection of information available on truck operation and maintenance.

Volume 1: Engines
JAMES E. BRUMBAUGH

*5 1/2 x 8 1/4 Hardcover 416 pp. 290 Illus.
ISBN: 0-672-23356-8 $16.95*

Volume 2: Engine Auxiliary Systems
JAMES E. BRUMBAUGH

*5 1/2 x 8 1/4 Hardcover 704 pp. 520 Illus.
ISBN: 0-672-23357-6 $16.95*

Volume 3: Transmissions, Steering, and Brakes
JAMES E. BRUMBAUGH

*5 1/2 x 8 1/4 Hardcover 1,024 pp. 905 Illus.
ISBN: 0-672-23406-8 $16.95*

DRAFTING

Industrial Drafting
JOHN D. BIES

*5 1/2 x 8 1/4 Hardcover 544 pp. Illus.
ISBN: 0-02-510610-4 $24.95*

Professional-level introductory guide for practicing drafters, engineers, managers, and technical workers in all industries who use or prepare working drawings.

Answers on Blueprint Reading
(Fourth Edition)
ROLAND PALMQUIST;
revised by THOMAS J. MORRISEY

*5 1/2 x 8 1/4 Hardcover 320 pp. 275 Illus.
ISBN: 0-8161-1704-7 $12.95*

Understanding blueprints of machines and tools, electrical systems, and architecture. Question and answer format.

HOBBIES

Complete Course in Stained Glass
PEPE MENDEZ

*8 1/2 x 11 Paperback 80 pp. 50 Illus.
ISBN: 0-672-23287-1 $8.95*

The tools, materials, and techniques of the art of working with stained glass.